D1796608

Guide to Energy Management

in the Built Environment

Publication Information

Published by The Institution of Engineering and Technology, London, United Kingdom

The Institution of Engineering and Technology is registered as a Charity in England & Wales (no. 211014) and Scotland (no. SC038698).

First published 2017

The Institution of Engineering and Technology,
Michael Faraday House,
Six Hills Way, Stevenage,
SG1 2AY, United Kingdom.

Copies of this publication may be obtained from:
The Institution of Engineering and Technology
PO Box 96, Stevenage, SG1 2SD, UK
Tel: +44 (0)1438 767328
Email: sales@theiet.org
www.electrical.theiet.org/books

While the publisher and contributors believe that the information and guidance given in this work is correct, all parties must rely upon their own skill and judgement when making use of it. Neither the publisher nor contributors assume any liability to anyone for any loss or damage caused by any error or omission in the work, whether such error or omission is the result of negligence or any other cause. Any and all such liability is disclaimed.

The moral rights of the authors to be identified as author of this work have been asserted by them in accordance with the Copyright, Designs and Patents Act 1988.

A list of organisations represented on this committee can be obtained on request to IET Standards. This publication does not purport to include all the necessary provisions of a contract. Users are responsible for its correct application. Compliance with the contents of this document cannot confer immunity from legal obligations.

It is the constant aim of the IET to improve the quality of our products and services. We should be grateful if anyone finding an inaccuracy or ambiguity while using this document would inform the IET Standards development team at IETStandardsStaff@theiet.org or the IET, Six Hills Way, Stevenage SG1 2AY, UK.

ISBN 978-1-78561-112-4 (paperback)
ISBN 978-1-78561-113-1 (electronic)

CONTENTS

Acknowledgements

The IET gratefully acknowledges the advice and assistance provided by the following people and organisations in the development of this Guide.

Lead technical author:
Cameron Steel CEng FIET MCIBSE FIHEEM MInstRE (BK Design Associates UK Ltd)

Additional contributors:
Vikas Ahuja (Imperial College Healthcare)
Prof Ian Bitterlin CEng FIET (BCS/techUK)
David Bleicher (Building Services Research and Information Association, BSRIA)
Andy Bolitho (British Retail Consortium)
Mervyn Bowden (Empirical Energy/Energy Institute, EI)
Valeria Branciforti PhD MArch BArch (formerly Knowledge Transfer Network, KTN)
Peter Brogan (The British Institute of Facilities Management, BIFM)
Allan Burns (Telemental)
Tom Choularton (BAE Systems Maritime Services)
John Cowburn CEng FIEE (Smart Energy Networks)
Martin Fry (City University, London / Energy Services and Technology Association, ESTA)
Bruno Gardner (Carbon Trust)
Jamie Goth (Scottish Futures Trust)
Paul Hasley CENv MEI LCEA MA (Cantab) MSc BSc (Hons) (Surrey County Council)
Luke Hedger BEng C.E.M (GlaxoSmithKline, GSK)
Keith Irwin (Bowers Projects)
Andrew Jones (Empirical Energy)
Blane Judd BEng FCGI CEng FIET FCIBSE (BLTK Consulting) Chairman
Keeran Jugdoyal (Atkins)
Russell Layberry (Environmental Change Institute, Oxford University)
Dr Andy Lewry DIC CEng FIMMM CEnv MSocEnv FEMA (Energy Managers Associations, EMA / Building Research Establishment, BRE)
Chris Little (LB Hounslow Council)
Paul Lowbridge (National Grid)
Mat Lown (Tuffin Ferraby Taylor LLP, TFT)
Gordon Ludlow (British Institute of Facilities Management, BIFM)
Craig Mellis (SALIX Finance)
Dexter Nicco (Westminster University)
Gareth Parkes (Energy Institute, EI)
David Sheen (British Beer and Pub Association)
Paul Smyth MBA BEng (Hons) CEng MIET (SALIX Finance)
Kris Szajdzicki BSc FEI (Energy Services and Technology Association, ESTA)
Brian Taylor MEI (Matrix Controls Solutions — a company of e.on)
Rachel Toresen-Owuor (Buckinghamshire County Council)
Sam Woodward MIET (Lutron Electronics)
Les Woolner (British Electrotechnical and Allied Manufacturers Association, BEAMA)
Bill Wright MA CEng FIET (Electrical Contractor's Association, ECA)

Appendix A2.2 contains diagrams that are based on a Plan of Work illustration originally featured in CIBSE/ADE CP1: Heat networks — Code of Practice for the UK.

The IET wishes to acknowledge and thank CIBSE, ADE and the originator, Mr Phil Jones CEng MSc FCIBSE MEI MASHRAE, for the kind permission to develop and reproduce the diagrams for use in this document.

Three case studies are shown in Appendices A1, A2 and A3 respectively. These studies first appeared in BRE Information Paper IP7/13.

The IET wishes to acknowledge and thank BRE and the original author, Dr Andy Lewry, for the kind permission to reproduce the case studies in this document and for other information that was also incorporated.

Introduction

1.1 Objective

The objective of this Guide is to provide knowledge and good practice guidance on the management of energy used in engineering systems within the built environment.

1.2 Aim

The Guide aims to provide clear and concise information on energy management that can be developed and applied to a number of different installations. There is no single solution: energy management must be specifically tailored to meet specific local requirements, but the principal aspects will apply to all installations.

The key is:

(a) to understand the requirements outlined within this Guide;
(b) set the requirements against the context of the reader's own installation and particular fiscal circumstances; and
(c) adapt the process to reduce the consumption of energy in a meaningful way.

1.3 Context — the politics of energy

Within the built environment, large amounts of energy are used as we all go about our daily lives whether that is at work, at rest or in our recreational activities. As populations increase, as nations develop, as technology advances, ever greater energy demands are placed on the world's natural resources.

Those charged within energy management must wrestle with the conflicting demands of using less energy and keeping the bills down, whilst maintaining a comfortable environment for people to carry on with their usual activities. Another challenge is allowing societies to develop in a sustainable way and for economic systems to be prosperous.

The politics of climate change are beyond the remit of this Guide. There is, however, increasing evidence that man's appetite for energy is at least partially responsible.

The need for us all to be smarter and more frugal with our consumption of energy is more than apparent on many levels, whether that is individually, as organisations, as industries, as nations or as a global society. What is clear is that the burning of fossil based fuels to provide energy is creating ever greater amounts of polluting gases and that is not good for the environment.

There is an immediate need for energy management systems that provide coherent strategies and practical guidance. Doing nothing and carrying on regardless (typically classified as 'business as usual') is not an option. Contemporary philosophies on energy consumption have to change.

There is a need to recognise that the balance between 'energy in', 'energy out' and 'energy wasted', whilst still interrelated, are part of a more complex series of connected parts of the energy equation.

Reducing energy wastage will involve a series of iterative steps and involve various approaches. Some will involve improving building fabric or insulation of pipework, some will seek to improve engineering technologies and controls, while some will need better processes and improvements in user behaviour.

A typical rule of thumb: an existing installation, with legacy equipment and no previous energy management guidance to users or processes, could provide up to 20 % of potential energy savings (reference Carbon Trust CTG 056) — less is more. Within that notional percentage, 15 % is often technically and economically achievable through modifying user behaviour, improving processes or replacing technologies, whilst a further 5 % could be more difficult to achieve and hence could be regarded as residual energy waste.

▼ **Figure 1.1**   Basic understanding of energy use

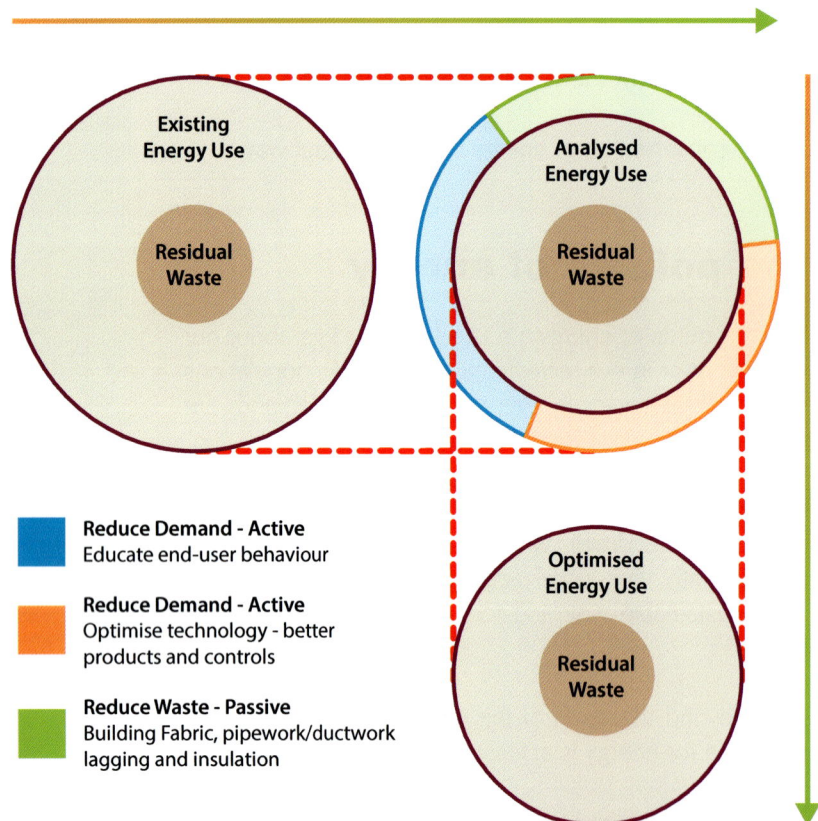

It should be recognised that there are links between energy management as a function of the business and its associated overhead costs. There are links to an organisation's responsibilities to carbon management and the associated responses to climate change legislation. Energy management also influences energy efficient infrastructure to assist in better use of energy.

It is worth noting though, that cleaner energy at any price can have unexpected consequences.

The conflicting demands of providing energy at a realistic cost for all users, whilst ensuring resilience and security of supply and also moving towards fossil free fuel sources for energy extraction, has been described as the 'energy trilemma' (illustrated in Figure 1.2).

This can be summarised as the challenge of keeping the lights on, at an affordable price, while decarbonising our power generation system.

▼ **Figure 1.2** Energy trilemma

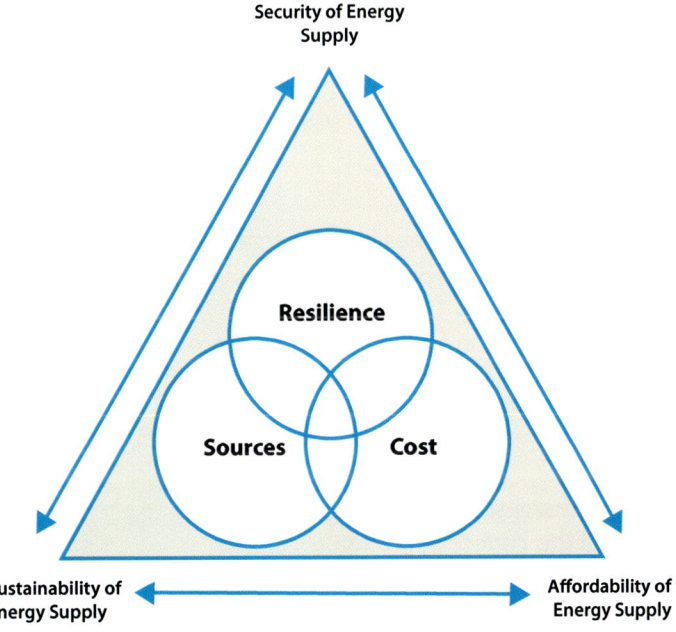

Finding the perfect balance between the three sides of the energy trilemma will remain the subject of much debate amongst the key policy stakeholders of national and international governmental bodies alike.

Where energy policy is concerned, there will also be a conflict between the short term expediency of electoral politics set against the backdrop of medium term policy of energy production and use. This in turn may be at odds with a long term strategy and international commitments, which are driving down the use of fossil fuels, with the aspiration for sustainable energy generation and usage.

1.4 Types of organisations/sites

The principles demonstrated in this Guide will apply to all commercial and industrial installations within the private and public sectors. The range of estates will vary from one simple building to large campuses with multiple buildings across one large site. Equally an estate may comprise of separate buildings in multiple locations.

1.5 Who should use the Guide?

This Guide should be:

(a) used by all those with specific or delegated responsibility for managing the procurement, consumption and control of energy;
(b) used to review and improve particular processes and procedures;

(c) understood by all engineering practitioners and managers to monitor the use of energy and plan associated improvements;

(d) used to target efficiencies in an effective way that supports the organisation's short, medium and long term goals; and

(e) reviewed by those with fiscal responsibilities to understand the technical challenges behind energy management and the resources required to ensure realistic outcomes and satisfactory procedures.

The role of an energy manager is developing all the time. Some organisations, typically larger estates and corporations, will have a clearly defined role for managing energy. Smaller organisations typically have the role as one more responsibility added to many.

1.6 Plan, Do, Check, Act

There are a number of accreditation and certification schemes for both organisations and individuals in the field of energy management.

This is often centred on BS EN ISO 50001, which is the international standard for energy management. These schemes ensure that:

(a) a business is assessing and reporting on their energy use, and using that data to improve the energy performance; and

(b) an individual has the correct training and skill sets to guide an organisation through the actions required and make the correct decisions.

BS EN ISO 50001 is based around the common four step cycle of 'Plan, Do, Check, Act' (PDCA) that is commonly used in health and safety, engineering maintenance and managing.

▼ **Figure 1.3** PDCA cycle

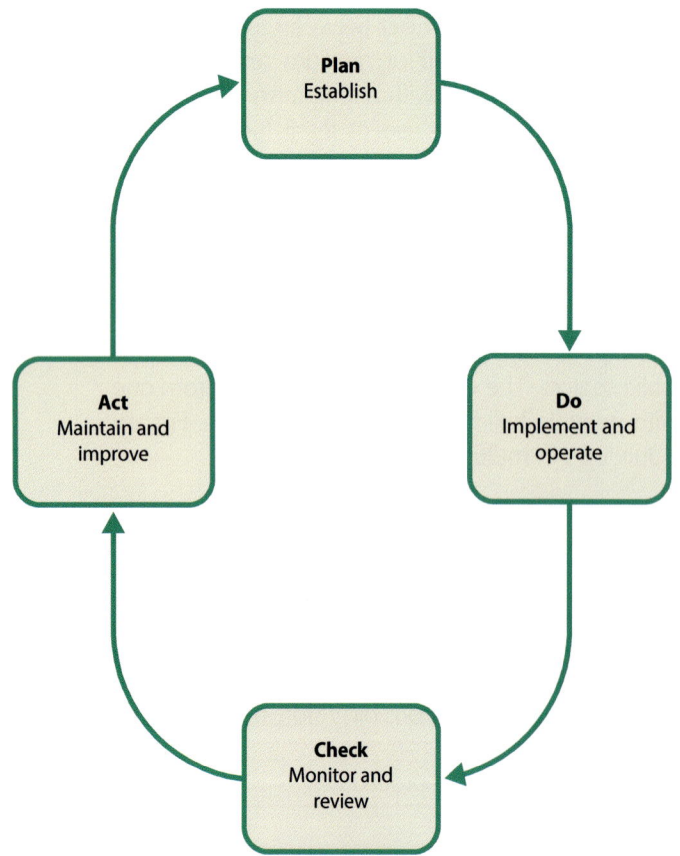

BS EN ISO 50001 describes a process that demonstrates an energy management system and the main operational headings within the document are shown in the table below. The 'Headline activity' column of the table below also shows how these headings align to the PDCA cycle.

ISO 50001		Headline activity
Section	Heading	
4.1	General requirements	
4.2	Management responsibility	Plan
4.3	Energy policy	Plan
4.4	Energy planning	Plan
4.5	Implementation and operation	Do
4.6	Checking	Check
4.7	Management review	Act

Figure 1.4 below further develops PDCA for ISO 50001 and incorporates links originally demonstrated in a BRE paper by Dr Andy Lewry. It shows the activities required to put into place an effective energy management system.

Further detail on the sub-headings for ISO 50001 can be found in Appendix D of this Guide.

▼ **Figure 1.4** PDCA links for ISO 50001

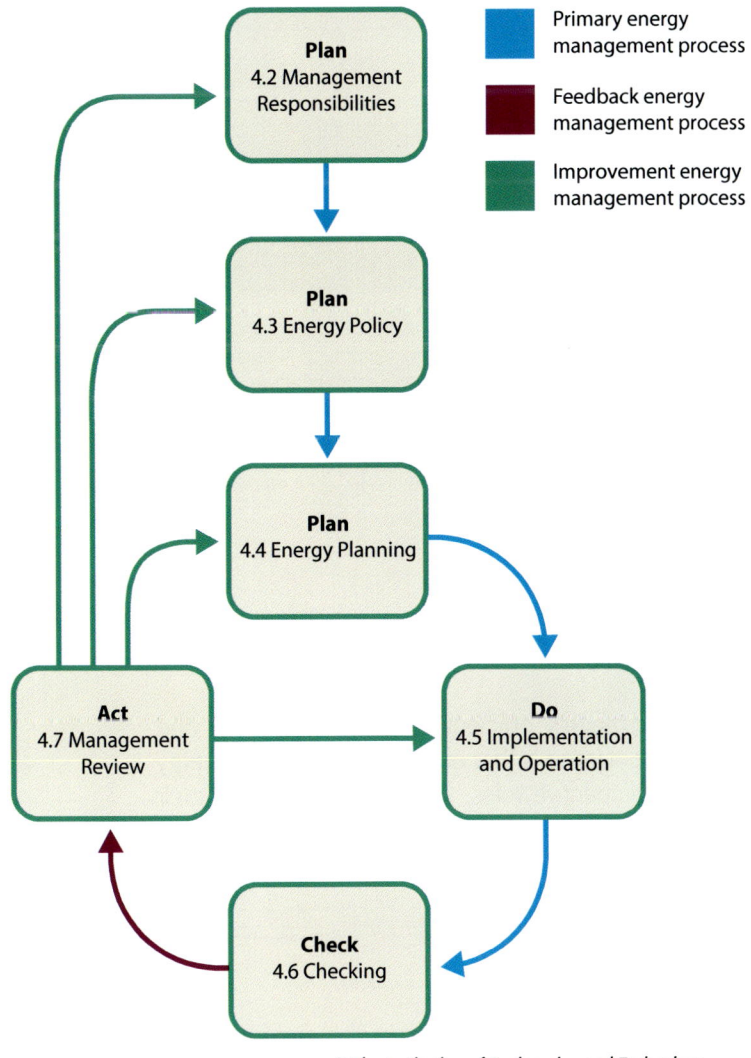

1.7 Energy efficient systems

Engineering systems, in the context of energy efficiency, can provide a slightly different approach to the four-step sequence. Here the emphasis could be to ensure that energy into the engineering system has sufficient capacity with minimum losses and is controlled satisfactorily to provide the correct amount of energy at the point and time of need. This activity should then be monitored to ensure the outcomes are correct, losses are accounted for and improvements assessed.

Another international standard IEC 60364-8-1 provides a diagram on the engineering system underpinning energy efficient electrical systems. This in turn can be presented in a simplified generic energy efficient schematic as shown in Figure 1.5 below.

▼ **Figure 1.5**   Energy efficient engineering cycle

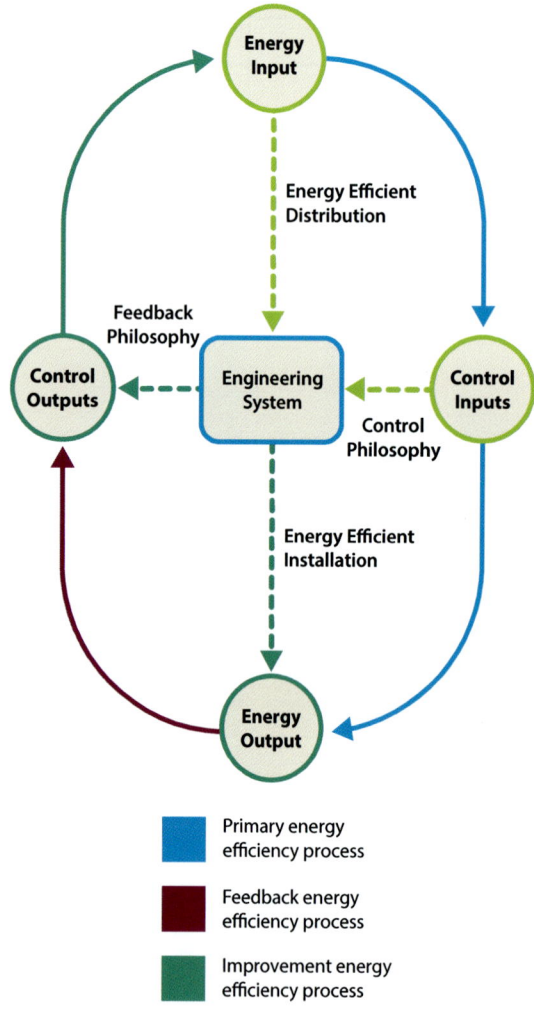

1.8 Link and coordination

This Guide demonstrates the link between energy efficient engineering systems and energy management systems. The desired outcome is that engineering staff and energy management teams can understand their respective roles, the overlaps between their respective responsibilities and how they need to coordinate and assist each other to improve the energy management of their particular installations.

1.9 Using the Guide

This Guide highlights the requirements of the pertinent national and international standards in this increasingly important area of energy management. It will build on these concepts and frameworks to provide practical examples and illustrations to those readers with a technical background. These can then be applied to energy management in a variety of installations types in and around the built environment.

For the non-technical reader, the Guide also seeks to provide information, guidance and confidence, so that when they are procuring expertise and energy management services from outside their own organisation, they can ensure they are getting the correct strategy and best value for money.

Links are provided to relevant documents, standards and associated publications to allow the reader to undertake a further detailed study of his or her own particular installation.

As an initial point of reference, this Guide will typically draw on other UK-centric documents. There will also be reference to major international standards where relevant. The reader should be aware that this approach is not intended to exclude other international or other national standards. Wherever the reader is located it will be possible to find equivalent or similar standards for the energy management systems that are quoted here.

The concepts of good energy management, and the basic principles that support that, remain the same worldwide and should still be applied following the intent described in this Guide.

The Guide is structured to provide good practice guidance on a number of areas within the energy management field. It is not intended to replace the accreditation schemes referred to above.

Section 2 of the Guide provides the principal explanation and supporting framework for energy management. Definitions are provided, concepts introduced and the links between energy efficient engineering infrastructure and energy management processes are explained. These links are set in the context of the PDCA cycle to give the reader an overview of what they need to achieve and an insight on how they can formulate their own tailored plan of action for better energy management of their particular organisation.

The four principal appendices that follow Section 2 provide greater detail around each particular stage of the PDCA cycle that supports and enhances those activities.

(a) Appendix A1 discusses how to plan: the requirements of energy policy, strategy and procedures, and the links to engineering design.
(b) Appendix A2 discusses what to do: the requirements of procurement, roles and competences and user behaviour, and the links to engineering maintenance.
(c) Appendix A3 discusses when to check: the requirements of performance checks, benchmarks and loss assessment.
(d) Appendix A4 discusses where to act: the requirements of reviewing targets, mitigations and improvement projects.

Throughout these four appendices there are a series of self-assessment questions (SAQs). These should be used by the energy management team to examine their particular organisation to improve their energy management system and processes.

Appendix B, as an energy management tool, collates all of the SAQs together for ease of reference. A scoring matrix is also provided as an indicative guide for energy managers and engineering staff so they can conduct a self-assessment process and evaluate where they are now. This evaluation can then be periodically reviewed to highlight improvements in each area.

The four supplementary appendices then provide cross referencing and further background information for the reader.

Appendix C provides an overview for energy managers into of some of the technologies, engineering considerations and design philosophies used in different building services engineering systems to make them more energy efficient.

Appendix D provides cross referencing and signposts to relevant standards and other associated documents that the reader may find useful.

Appendix E provides an overview on different types of energy models, based on both traditional energy infrastructure and more contemporary routes to energy procurement and consumption.

Appendix F explores a future change of mindset and approach to energy management and energy conservation, through an introduction to the seven steps of transition engineering. This provides tools that seek to significantly alter short to medium term philosophies of energy management, and existing change management approaches, to something with a long-term philosophy that is more holistic and ecological in outlook.

Transition engineering also approaches the problem, not just from an engineering perspective, but also taking into account overlapping aspects of environmental science, social science and economics.

SMART energy management

Strategy and policy defined

Managed energy consumption and procurement

Actual performance measured and monitored

Reviewed against benchmarks and alterations

Targeted improvements and changes

▼ **Figure 1.6** Graphical overview of contents

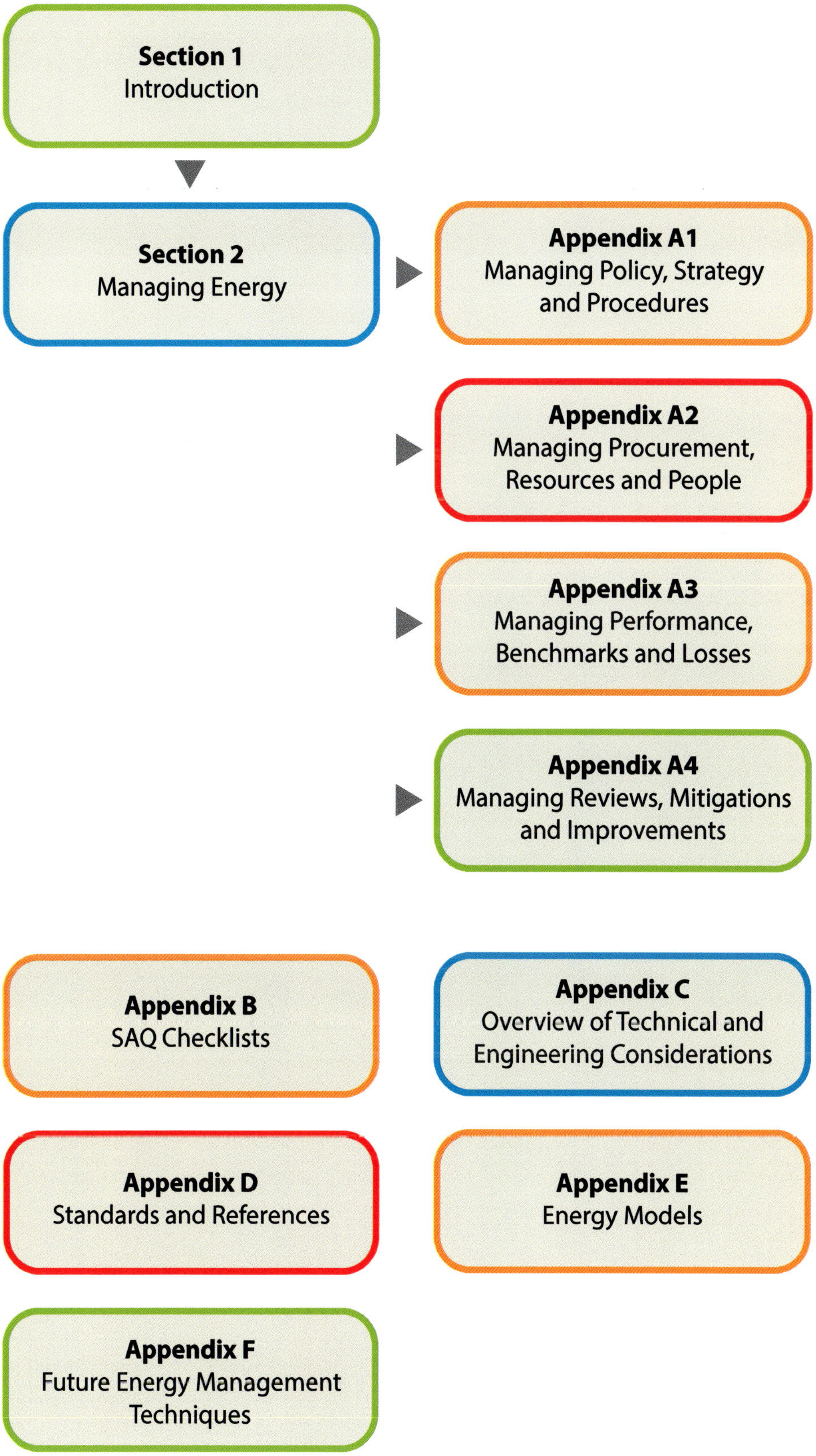

Section 1
Introduction

Section 2
Managing Energy

Appendix A1
Managing Policy, Strategy and Procedures

Appendix A2
Managing Procurement, Resources and People

Appendix A3
Managing Performance, Benchmarks and Losses

Appendix A4
Managing Reviews, Mitigations and Improvements

Appendix B
SAQ Checklists

Appendix C
Overview of Technical and Engineering Considerations

Appendix D
Standards and References

Appendix E
Energy Models

Appendix F
Future Energy Management Techniques

SECTION 2

Managing energy

2.1 Describing an energy system in the built environment

A built environment energy system is any process within a building, or its immediate surroundings, that uses an energy input, typically from electricity, gas or fuel oils, to provide a service either to improve, or maintain, an operational environment or to enable products to be manufactured.

Examples of energy use in buildings and industry will include the provision of:

(a) heating, ventilation and air conditioning;
(b) water heating;
(c) interior and exterior lighting;
(d) internal transportation (lifts and escalators);
(e) telecommunication;
(f) data processing and storage;
(g) security and life safety systems; and
(h) industrial processes, for example, mining, agriculture and chemical works.

An energy system will consist of an energy input, some losses within one or more energy conversion processes, controls or user interfaces and a resultant energy output.

2.2 Defining an energy management system

From BS EN ISO 50001, an energy management system is defined as a:

"set of interrelated or interacting elements to establish an energy policy and energy objectives, and processes and procedures to achieve those objectives".

2.3 Commentary on energy management systems

An energy management system can be further described as a series of interconnected processes by which the distributed use and consumption of energy is made more efficient. This will involve the use of policy, strategy, action plan, control and monitoring tools to:

(a) benefit the environment by decreasing emissions;
(b) improve the operation of an installation by reducing wastage;
(c) improve the financial side of an organisation by reducing energy costs as an overhead;
(d) set targets to reduce wasted energy;
(e) implement procedures to maintain and continually improve efficiency; and
(f) ensure compliance with the legislative requirements on reporting of energy usage.

© The Institution of Engineering and Technology

An efficient energy management system will seek to call on the right amount of energy input, at the most cost effective time, with minimal conversion losses, coupled with optimal controls and subsequent minimal energy output to achieve the desired result.

The system will include activities to monitor energy usage and compare the results with predetermined targets and with previous usage. Where appropriate an energy management system will seek improvements in use and efficiency.

Within the built environment energy management systems should aim to reduce energy consumption through improved designs, use of better products, careful system operations and efficient building envelopes.

The outcome of these improvements should not conflict with the requirements of:

(a) health and safety legislation and best practice;
(b) environmental protection legislation and best practice;
(c) the primary focus of core business activities;
(d) the outcome of these improvements should enhance and embed a culture of energy management both within the management and throughout the organisation;
(e) maintain the built environment at comfortable levels for the end users;
(f) maintain and improve productivity;
(g) be financially planned and incorporated into the larger resources programme of an organisation; and
(h) overall, make better use of natural resources, whilst being kinder to the wider global environment.

Good practice resulting from carefully planned and monitored installations will benefit the end user and potentially extend the lifecycle of the equipment that has been deployed.

2.4 Energy management and the installation lifecycle

The energy management process should be a controlled sequence of events and activity to demonstrate a clear and transparent record of improvements and compliance with national and international standards.

Within the built environment, the energy management perspective should pervade all aspects of the life cycle of a product, a building, a transport system (such as lifts or escalators) or an industrial process.

Key areas of consideration to ensure use of best practice industry guidance are:

(a) careful design that considers the whole life cycle;
(b) compliant procurement and installation;
(c) satisfactory commissioning, handover and training processes;
(d) robust operation and maintenance management procedures;
(e) potential recycling or reuse of materials or equipment; and
(f) subsequent and careful disposal where necessary.

The use of guidance, including sustainability methodologies such as Building Research Establishment Environmental Assessment Methodology (BREEAM) and Leadership in Energy and Environmental Design (LEED), can help inform the design team and provide benchmarks for the design of energy efficient buildings and associated services installations. Other methodologies, such as CEEQUAL, cover carbon reduction for developments and infrastructure installation.

The supply chain should be carefully assessed to ensure it too subscribes to an energy management agenda. This should be demonstrated through the use of best practice procurement standards.

Other guidance and design criteria for particular building services disciplines should also be consulted, for example:

(a) IEC 60364 *Low-voltage electrical installations — Part 8-1: Energy efficiency*
(b) IET *Design Guide for Energy Efficient Electrical Installations*
(c) IET *Guide to Metering Systems: Specification, Installation and Use*
(d) CIBSE Guide F *Energy Efficiency in Buildings*
(e) CIBSE Technical Memorandum 39 *Building Energy Metering*
(f) CIBSE Technical Memorandum 46 *Energy Benchmarks*
(g) CIBSE Technical Memorandum 54 *Evaluating Operational Energy Performance of Buildings at the Design Stage*
(h) CIBSE Technical Memorandum 56: *Resource Efficiency of Building Services*
(i) ASHRAE Advanced Energy Design Guides
(j) BREEAM Building Research Establishment Environmental Assessment Method
(k) CEEQUAL Civil Engineering Environmental Quality Assessment
(l) LEED Leadership in Energy and Environmental Design

Further directives, standards, guidance, and papers from other organisations, are referenced in Appendix D of this Guide.

The energy manager's role as a stakeholder in the design of engineering systems and their subsequent operation should not be overlooked.

2.5 Schematic representation of an energy system

Energy managers should recognise the key elements required for an energy efficient design and installation. As a key stakeholder in any design stage they should be aware of the systems specified and how they will affect the ongoing operation of their site (refer to Figure 2.1).

2.5.1 Energy supply

Energy supplied to a site, whether from electricity, gas, oil, biomass or other sources will fall into one of three main categories:

(a) grid connections: typically to mains electricity and mains gas and sometimes to district heating schemes.
(b) local renewable energy: typically, electrical from photovoltaic or wind turbines or heat from solar thermal systems, or more occasionally, geothermal. Other forms of local energy production could include combined heat and power (CHP) although this is typically connected to mains gas as the primary energy source.
(c) energy storage: typically oil tanks or bio mass for off-grid boilers, occasionally local water reservoirs connected to electrical turbines, thermal heat store technology typically connected to air source or ground source heat pumps (e.g. hot tar and hot water) and newer technology involving battery storage systems connected to photovoltaic sources or fuel cells.

2.5.2 Control of energy inputs

Energy system controls usually fall into three main categories:

(a) Environmental sensors that measure against defined set points and provide signals to indicate when particular equipment should provide their service. Examples of this could be the automatic operation of:

   i   lighting when lux sensors indicate low or high levels of natural light;

   ii   heating when thermostats indicate low or high temperature;

   iii   extract fans when humidity sensors indicating high moisture in the room environment; and

   iv CO_2 sensors activating ventilation when air quality falls below certain levels.

(b) Energy availability that assesses what energy sources are available at the particular time. For instance, if local generation is being used and fails then the grid-connected supply will be required as back up.

   In more advanced systems the pricing for energy from the grid at a particular time of day may also be assessed to control costs as much as energy demands.

(c) User requirements: it is important to automate the control of energy use as much as possible, but local user requirements should always be considered first. In some environments the needs of users or critical equipment may override energy efficiency, examples of this will be:

   i   the health and wellbeing of patients in hospitals;

   ii   cooling requirements and electrical supplies in data centres; and

   iii   precision production equipment in industrial premises.

2.5.3 Energy system loads

Installation loads should be categorised according to their importance and when they might be used. The simplest form of energy management is simply to turn a load off; however doing so may not be desirable in all cases.

The concept of load inertia describes the load's reaction to be first being switched off to shed load and subsequently being reinstated. There may be consequences and effects on the levels of service.

A high inertia load, in simple energy management terms, may be an electrically powered immersion heated water tank. Short term loss of power to a tank that is already up to operating temperature may not have an immediate consequence on the end users.

Some larger industrial loads could be considered low inertia. Loss of power during the middle of a set manufacturing process could result in imperfections to the product and potentially give rise to consequent safety issues. Milling machines with fine tolerance settings are one example of this. A loss of power will result in loss of materials and associated costs. Correct analysis of the infrastructure design could mitigate these potential problems.

A low inertia load could also be particular types of lighting technology which must cool down sufficiently before being switched back on. The length of time for the lighting to be restored would be considered inconvenient and the loss of light may be possibly dangerous to occupants.

Lighting in public spaces could be a low inertia load. Loss of power to lighting in these areas could have consequences on the level of service to end users and their ability to carry on with their activities. Emergency lighting would be provided as a statutory requirement but this would be set at levels to enable safe evacuation.

An off-peak load would be categorised by a load controlled by a timer that takes advantage of a lower tariff. A traditional example of this would be electrically powered storage heaters.

2.5.4 Utilisation of control output

The controls philosophy of an energy efficient installation should be designed to provide a number of outputs that enable the user to monitor performance and make informed decisions to improve those operations. Three key outputs would be:

(a) trends and monitoring: this enables an energy manager to highlight medium to long term energy consumption and assess remedial activities with users in the area;

(b) energy billing: this enables an energy manager to ensure correct billing from the utility supplier and also billing for sub-tenants on site or internal cost centres; and

(c) user guidance or alerts: using controls provide short term alerts if energy use is not controlled or reaching unacceptable levels.

▼ **Figure 2.1** Schematic representation: engineering design for energy efficiency

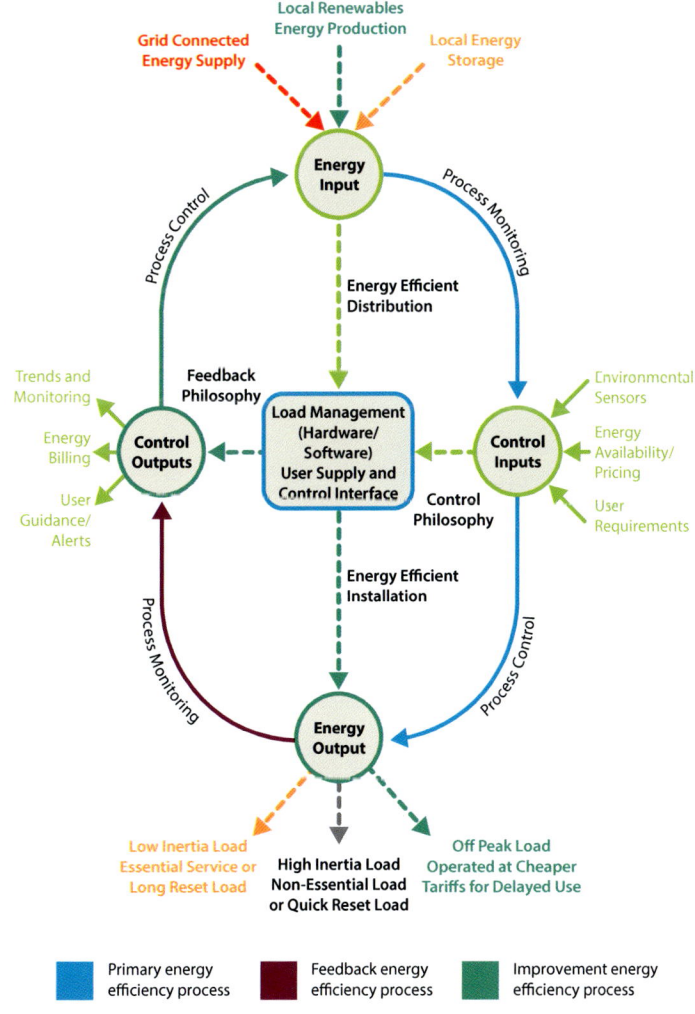

2.6 The interface between engineering design and energy management

The key activities of an energy manager are represented below and will fall into four main areas – purpose and principles that proactively lead the management of the energy agenda, whilst parameters and progression reactively manage outputs and planning. The activities should be seen as a cycle of events that are continuously being reviewed and seeking improvements.

▼ **Figure 2.2** Energy management interface

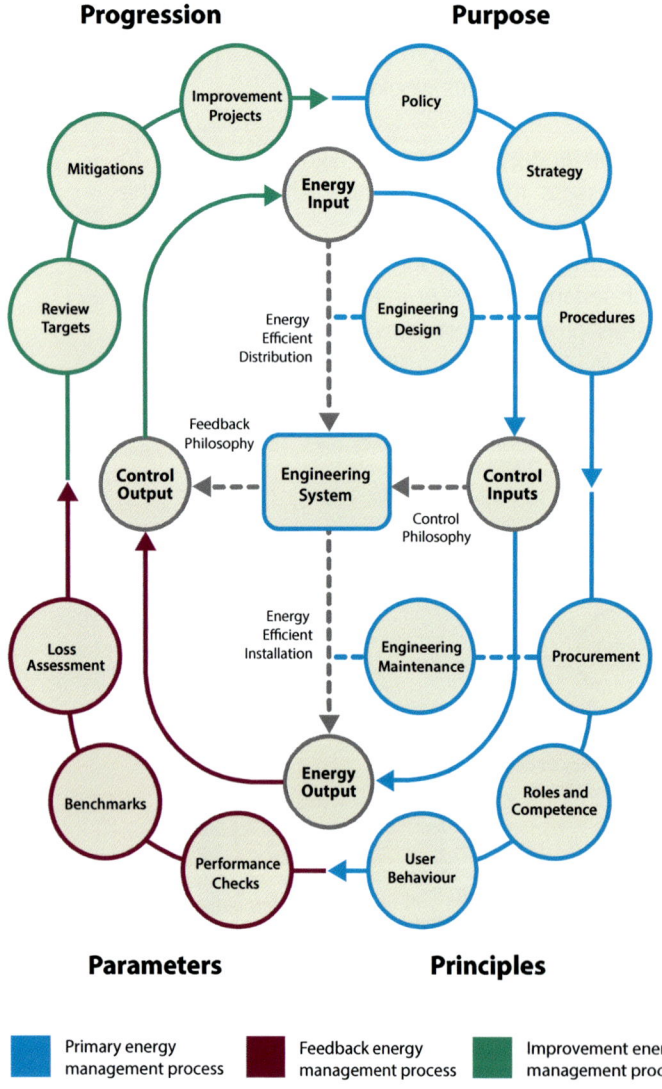

Purpose	Plan

All organisations, at a senior level, should recognise that proactive groundwork for energy management is laid by understanding and defining its purpose properly.

Working with the organisation management, the energy manager should develop a strong energy policy for adoption at director level that:

(a) encourages better use of energy;
(b) reduces dependency on fossil fuels where appropriate; and
(c) adopts use of sustainable energy sources where feasible.

A robust energy strategy should be defined that:

(a) understands the existing energy requirements of all cost centres within an organisation;
(b) provides clear targets (key performance indicators) with defined milestones;
(c) accounts for improvement projects in energy usage;
(d) accounts for risks and barriers to improve in energy management;
(e) understands future business strategies;
(f) analyses risk to energy capacity and energy resilience; and
(g) is underpinned with SMART objectives (refer to Section 1).

Clear energy procedures should be implemented that show how:

(a) energy use will be managed;
(b) energy intake will be procured;
(c) energy consumption will be monitored;
(d) energy loss will be mitigated;
(e) energy targets will be set; and
(f) energy performance will be reviewed and improved.

The organisation should also understand what input energy it requires and what aspirations it has to the use of local renewable technology and also any resilience issues associated with using local storage.

Energy managers should be seen as important stakeholders in any engineering design process to ensure efficient distribution and control of energy around the site. Energy efficient designs will also inform site energy procedures. New equipment will require training of site resources or engagement with contractors to ensure efficient operation.

Section 2.9 below provides general advice on the relevant elements of both energy efficient systems and this part of the energy management cycle.

Within this Guide, Appendix A1 provides greater detail on the purpose of energy management, together with requirements, tools and relevant cross references to other documents.

Principles	Do

At an operational level, proactive principles of energy management define specific requirements covering particular areas.

Procurement of energy from external sources will be driven by cost, logistics and ethics so that:

(a) there is a clear understanding of the existing energy procurement contracts;
(b) the types of energy grid connections and peak capacity are known;
(c) resilience strategies are understood;
(d) the requirements of off-grid fuel supply deliveries are understood; and
(e) business appetites for renewable technologies are explored, both on and off-site.

Other aspects of procurement will include an organisation's agreed standards for energy efficient and near zero carbon technologies.

There should be a clear recognition of the roles and responsibilities of all personnel and visitors with respect to daily energy management so that:

(a) all directly employed and temporary staff understand their own responsibilities on energy use;

(b) all visitors are guided on the expectations of the host organisation;

(c) all Facilities Management and Estates staff understand their additional roles regarding the upkeep, housekeeping and maintenance of energy efficient systems and equipment;

(d) all Estates staff are briefed on the importance of energy management when developing estate strategies and the acquisition/disposal of property;

(e) training and knowledge dissemination is provided where required; and

(f) a skills matrix is circulated to help clarify requirements and identity skills gaps on energy management issues.

A clear and coherent programme to inform and encourage user behaviour so that:

(a) energy efficient operations are the new norm and not 'someone else's problem';

(b) low cost technology solutions can be rolled out across the organisation to enable simple energy waste reductions measures;

(c) users are properly educated on any installed high technology controls for energy management purposes and the risks to energy consumption levels if those controls are overridden; and

(d) future energy saving opportunities can be identified for further investigation.

Energy managers should closely liaise with:

(a) the commissioning and handover processes of new installations;

(b) the operation and maintenance of existing installations; and

(c) the end users to ensure that working environments and, hence, productivity are not compromised.

Any energy efficient design that is not set up and commissioned properly and correctly will quickly be subjected to the performance gap issues that are often encountered in the built environment. There is also evidence, from various sources, that poor ventilation in a building will affect the internal air quality, the health and wellbeing of occupants and associated productivity.

Similarly, any energy efficient design that is not maintained correctly will not save energy. Careful scheduling of any mechanical or electrical installation maintenance will ensure that the optimum energy performance of plant and equipment is provided as much as is reasonably practicable.

Section 2.10 below provides general advice on the relevant elements of both energy efficient systems and this part of the energy management cycle.

Within this Guide, Appendix A2 provides greater detail on the principles of energy management, together with requirements, tools and relevant cross references to other documents.

Parameters	Check

Auditing and reviewing of the operational parameters of energy management will follow a reactive approach. The energy manager should be aware of the key monitoring activities that are required to measure, monitor and assess energy performance.

Regular performance checks from strategically placed energy meters will enable timely and accurate analysis so that:

(a) normal energy patterns of use can be monitored;

(b) abnormal energy events can be highlighted for further assessment;

(c) power quality assessments on electrical infrastructure can be monitored; and

(d) billing of tenants with appropriate meters can be made.

The energy manager should be aware of the expected performance benchmarks for the building:

(a) for new buildings this will be the design data and the initial commissioning measures on practical completion;

(b) for older buildings the initial design data should be updated with seasonal data in subsequent years for direct comparison; and

(c) recent projects and developments will also inform and update the site wide energy performance parameters.

The energy manager should carefully compare both the performance check data and the known benchmarks to reconcile and evaluate any loss assessment so that:

(a) they can demonstrate that the information from both data sets is consistent. Careful use of recognised mathematical tools should be considered, such as regression analysis and cumulative sum (CUSUM) control charts amongst other methods.

(b) they can be sure the information accurately illustrates evolving energy use in the building. No two buildings are alike and their particular use often changes as sections are either upgraded or fall into disuse.

(c) they can highlight and account for any adverse events. Comparison with key performance indicators and complemented by exception reports will highlight wasted energy or additional use for particular activities.

Section 2.11 below provides general advice on the relevant elements of both energy efficient systems and this part of the energy management cycle.

Within this Guide, Appendix A3 provides greater detail on the parameters of energy management, together with requirements, tools and relevant cross references to other documents.

Progression	Act

Building on the reactive philosophy, the progression of energy management should provide a comprehensive analysis of recent energy usage and sets the next cycle of energy management in motion.

The energy manager should periodically review targets to ensure that:

(a) the existing targets match the current capacity of the installation;

(b) targets are revised down for installations that are reducing in size;

(c) targets are revised upwards for installations that are expanding in size; and

(d) any new developments are catered for and the timescales for implementation are understood.

Any reviews should take into account any changes in use, occupancy hours and occupant density in particular areas. Energy managers should be wary that any targets are not normally absolute or that one size fits all; bearing in mind that installations are invariably complex.

Energy performance indicators (EPI) such as kWh/m2 could go down as organisations expand due to economy of size. Targets need to be specifically tailored to the operation and aims of any organisation; therefore, they need to be granular and not just catch all.

The energy manager should consider any mitigations so that:

(a) it is understood which parts of the organisation are critical to business success and may inhibited by overzealous energy efficiency measures;

(b) it is understood where any barriers to improvements in energy use exist, from user behaviour to supply infrastructure to system controls; and

(c) they can accurately inform any regular review of the business energy policy, energy strategy or associated procedures.

The energy manager should be a key stakeholder in any improvement projects with input in any design, preparation, procurement and implementation processes so that:

(a) the proposals concur with overall energy policy and planned strategy;

(b) any necessary changes to energy procedures are flagged up early on;

(c) any additional operational resources are identified early on;

(d) proposed energy consumption forms part of any business case;

(e) retrospective improvements to existing facilities comply with site energy strategy; and

(f) new energy efficient equipment complies with best practice procurement standards.

Section 2.12 below provides general advice on the relevant elements of both energy efficient systems and this part of the energy management cycle.

Within this Guide, Appendix A4 provides greater detail on the progression of energy management, together with requirements, tools and relevant cross references to other documents.

2.7 Coordination between engineering and energy management

The following tables demonstrate the roles that engineering design and energy management play in achieving energy efficiency. There are parallels and links between the two disciplines, as shown in Figure 2.2, and also overlaps with maintenance.

It is vital that the disciplines work in a coherent and coordinated manner to:

(a) enhance the installation design;

(b) improve its operational efficiency; and

(c) obtain full value from the life cycle of the equipment installed.

2.8 Purpose of an energy management system

Item	Engineering design model	Energy management model	Comments
1		Policy	Adopted by the organisation senior management, complete with a board room champion and sign off by the board executive, an energy policy defines the organisation's approach to energy use for business needs. The energy manager should advise on the creation of the policy and administer periodic reviews and updates. Further guidance in Appendix A1.1.
2		Strategy	Describes how the policy will be implemented within the organisation and how it will leave the organisation with growth capability. The energy manager should review annually, consult with the senior management, including capital projects and facilities managers, and adjust according to the perceived future needs over appropriate periods. Further guidance in Appendix A1.2
3		Procedures	Demonstrates particular methodologies and how energy will be managed, monitored and used efficiently. Headline design criteria can also be specified with targets for use and efficiency bands for equipment defined to international standards. Further guidance in Appendix A1.3.
4	Energy input		The actual type of energy source(s) to a site may be constrained by engineering practicalities or potentially influenced by energy policy. For instance, whilst the aspirations may be for renewable supplies, the site may not allow for installation of a wind turbine. Can the policy be fulfilled by participating in a community level renewable energy scheme, or at least signing up to a green electrical tariff?
5	Grid-connected		The capacity of the site energy grid connection should be carefully assessed, accounting for current known requirements and any foreseeable future needs too. Underestimating will inhibit future growth. Overestimating will result in overpaying for ongoing excess capacity charges.

Item	Engineering design model	Energy management model	Comments
6	Local energy production		The use of local energy production should be carefully assessed. As well as initial capital expenditure, time and costs for labour and materials for routine maintenance, emergency repairs and associated downtime should be considered, with the impact on any grid capacity. Resilience and associated dependency on supplies for the installation should be carefully assessed.
7	Local energy storage		Newer technology will allow for energy storage derived from renewable sources. Maintenance requirements for this should also be considered. Contingency plans for downtime will also need to be developed.
8	Energy efficient distribution	Engineering design	An overlap between the energy manager and engineering designer ensures that the design meets the site strategy. Building services infrastructure should be designed and constructed to be energy efficient, whilst also being fit for purpose to allow sustained operation and potential growth in the business. The supply point should be as close to the point of use as possible. The energy source should be metered and where appropriate sub-meters used throughout the installation. Equipment at the point of use should be energy efficient and with the correct design control philosophy. Further guidance in Appendix A1.4.

2.9 Principles of an energy management system

Item	Engineering design model	Energy management model	Comments
9		Energy procurement	An energy manager should periodically review all energy costs and associated tariffs. This information needs to be matched against the existing and future needs of the organisation. Electrical supplies in particular may be derived from off-site renewable sources to match with the energy policy. Further guidance in Appendix A2.1.
10		Roles and competence	Energy managers should assist with the definition of roles and responsibilities throughout the organisation and delegate key tasks to assist with day to day management of energy. A matrix of responsibilities will assist with defining these roles and the required training for personnel with key tasks throughout the energy management spectrum. Further guidance in Appendix A2.2.
11		User behaviour	Energy loss is a risk to business continuity. Energy managers should lead the analysis of user behaviour and seek to influence that behaviour by education, specific training and, where absolutely necessary, restrictive use. Site inductions of energy management procedures, regular communication of energy management updates and periodic training of key stakeholders will assist in this. Further guidance in Appendix A2.3.
12	Control inputs		Control inputs provide information on the parameters and operating environment within which the energy use takes place. Some will be automatic and trigger signals at set points, others will form part of the operational specification for the space.
13	Environmental sensors		Engineering control philosophy must include sensors to report on the local ambient conditions. Set points will signal whether an engineering system should be switched on or off. Simple internal sensors might be thermostats for space heating or water heating. Response to external factors, for instance lux sensors, will turn off luminaires when there is sufficient natural light.

Item	Engineering design model	Energy management model	Comments
14	Energy availability/tariffs		Engineering controls required for off-peak or lower energy tariffs for some large loads. Engineering design should also be considered to allow for separation of infrastructure too. Energy storage may not have the capacity required in the short term, requiring connection to the grid in the interim.
15	User requirements		Defined parameters, regulations and guides will provide a base design. It is important to provide local controls that allow users to adjust the set points to provide comfortable temperatures or override automated lighting.
16	Energy efficient installation	Engineering maintenance	An energy efficient design needs to be converted successfully into an energy efficient installation. The installation then needs comprehensive commissioning and robust maintenance to remain energy efficient throughout its working life. Energy managers need to be stakeholders through: (a) the design process so that they understand what is being provided; (b) the commissioning process so they understand what the system has achieved and how it will operate; and (c) the maintenance process to ensure the installation continues to deliver. Maintenance teams should be competent and trained to ensure correct operation of energy efficient equipment. Maintenance regimes should be planned to ensure satisfactory levels of preventative maintenance. There should be contingencies in place for reactive maintenance to deal with system failures. Further guidance in Appendix A2.4.

2.10 Parameters of an energy management system

Item	Engineering design model	Energy management model	Comments
17	Energy output		Energy output should be monitored and measured where practical so that energy use can be assessed and improvements implemented if necessary.
18	Low inertia load		Critical loads are low inertia: isolation via automated energy controls may cause detrimental service to people and property. Loads with poor reinstatement procedures will potentially waste more energy restarting than if kept running.
19	High inertia load		Non-critical loads, which may have some form of integral energy storage, will be high inertia. Loss of service for short periods will not affect the delivery of service. Loads that are quickly reinstated with minimal loss of energy are also high inertia. Isolating these loads when not required will assist with energy profiles.
20	Off-peak load		Loads that can be used during periods of off-peak tariffs. Service may include some form of mechanical energy storage for delivery at peak hours or an intensive energy consumption process that has been delayed to take advantage of a lower tariff.
21		Performance checks	Energy managers should regularly check the performance of the building services systems so that they understand what energy is being used, where and when. This should be undertaken using recognised metering, monitoring and targeting (MM&T) techniques. As stakeholders in the design process energy managers should influence the deployment of energy meters in appropriate locations to ensure the correct information is available. Further guidance in Appendix A3.1.

Item	Engineering design model	Energy management model	Comments
22		Benchmarks	Energy managers should request that the building's commissioning and handover documents, including base design and O&M manuals, provide clear guidance on the expected performance outcomes. Residual design risk assessments may also provide insights for any subsequent underperformance evaluation. As operations data is collated over a period of time, and as the building use evolves, it is important that these benchmarks are reviewed and, if necessary, updated. Further guidance in Appendix A3.2.
23		Loss assessments	As part of the MM&T activities, energy managers should reconcile actual performance with the base design. Additional comparisons can be built up year on year. Any underperformance should be carefully assessed and reports compiled so that mitigations can be put in place. Further guidance in Appendix A3.3.

2.11 Progression of an energy management system

Item	Engineering design model	Energy management model	Comments
24	Control output		Control systems should provide a series of accessible outputs to enable detailed analysis to be undertaken by the energy manager and the process operator. This allows real time tuning of the system, long term planning for improvements and, potentially, fiscal reconciliation.
25	Trends and monitoring		To assist the energy manager in MM&T processes, designers should ensure that control systems provide facilities to monitor trends in energy use, via metering, sub-metering, and outputs from the environmental sensors. Consultation with end users will ensure that the correct user interfaces are provided and the correct parameters are measured to monitor operational energy efficiency.
26	Energy billing		On large estates or multiple occupancy buildings metering will be provided for energy billing. Care should be taken on specifying the correct regulated and calibrated meters for this purpose.

Item	Engineering design model	Energy management model	Comments
27	User guidance/ alerts		Monitoring of energy performance in real-time is important to both the energy manager and the operations manager. To assist with engineering operations, comprehensive building energy management systems can be configured to provide alerts to facilities staff when defined operational conditions are breached to warn of energy usage patterns that fall outside of the norm. Predetermined control strategies will define upper and lower set points and also mid points to warn of potential issues.
28		Review targets	The energy manager should have access to and review all of the energy performance benchmarks continually. The data should include the original design information, commissioning data and also updates that demonstrate patterns of use in years since commissioning. This should be compared to actual usage and trends examined carefully. Further guidance in Appendix A4.1.
29		Mitigations	The energy manager should examine the trends to establish the reasons behind current performance. Has the occupancy profile changed? (a) If there has been growth in site wide activities, are the parameters now unrealistic? (b) If there is a decline in site wide activities, are the parameters encouraging waste? Has the energy manager been involved in business planning? (a) Does the energy manager have enough information to plan ahead? (b) Will energy procurement be a reactive exercise? Further guidance in Appendix A4.2.
30		Improvements	Following a review of existing consumption, the energy manager should identify actions, and therefore projects, to optimise energy consumption. These actions and projects may mean: (a) maintaining current good practices; (b) reminding and educating users on their obligations; (c) highlighting deficiencies and plans to remedy; (d) highlighting improvements in project procurement policies; (e) seeking better tariffs; (f) influencing new capital project designs; and (g) retrofitting of energy efficient equipment. Further guidance in Appendix A4.3.

2.12 Summary

For any energy management system to work satisfactorily it needs commitment and the full involvement of everybody within an organisation and its associated premises including staff and visitors. Whilst many larger organisations will have dedicated energy managers, many more do not – the role will be delegated to somebody who already undertakes many other duties.

The role of an energy manager is evolving. Ultimately, though, they are just a single point of contact and are simply providing the framework to guide the ship when it comes to reducing the consumption of energy, whilst maintaining the environment required by the occupier.

Everybody in the organisation, at every level, will have responsibility for their own actions on energy management in much the same way that health and safety, and environmental management, also carries duties. It is important that those roles and responsibilities are fully understood and knowledge about them is disseminated satisfactorily.

This Guide provides advice, tools and references to improve energy management processes in the built environment.

It follows the lead of international standards including BS EN ISO 50001 structures around PDCA.

It also illustrates the courses of action required for successful implementation of improved energy management processes.

The links to engineering design and management guide the reader to a wider picture.

It is important that the energy manager does not feel isolated in guiding an organisation to better energy consumption. They need to:

(a) be supported correctly by board level decisions and provided with the correct level of authority to carry out their respective duties successfully; and
(b) coordinate with others to ensure that the energy management processes work with:
    i   the commercial needs of the business; and
    ii the designs, commissioning and operations of the engineering part of the business.

Managing policy, strategy and procedures

A1.1 Energy management policy

Having an energy policy is an opportunity to demonstrate that the organisation takes energy management seriously.

It is vital that an energy policy document is owned, authorised and signed off by the senior leadership team in the organisation in the same way that environmental, health and safety, and equal opportunities policies are also signed off and that it carries the same level of importance.

Annex B of BS EN ISO 50001 provides a table demonstrating the consistency between various management process standards such as ISO 9001, ISO 140001 and ISO 50001.

The policy should be also subjected to annual reviews to ensure it is for purpose and updated as the needs of the business develop.

▼ **Table A1.1**   Energy policy self-assessment questionnaire

	Self-assessment question: What is the Energy Policy status?	Yes, or No?
a	The organisation has no energy policy.	
b	There is an energy policy, but it is incomplete, out of date or not signed off by the organisation's senior leadership.	
c	There is an energy policy in place that is accepted by the organisation's senior leadership, periodically reviewed with responsibilities delegated satisfactorily.	

An energy policy demonstrates that the organisation:

(a) is committed to managing energy consumption;
(b) is complying with legislation, recognised standards and criteria;
(c) understands associated industry good practice;
(d) encourages energy efficiency throughout the organisation;
(e) understands the costs of energy use and losses;
(f) has a framework to implement improvements and regular updated action plans on energy management; and
(g) understands global and national energy reduction targets and the timelines of any associated legislation to frame the development of energy strategy.

The organisation's senior leadership should provide a robust energy policy that gives a clear set of commitments and a clear set of objectives. This guidance will create a working environment that allows the energy manager to adopt strategies to:

(a) encourage better use of energy;

(b) reduce dependency on fossil fuels where appropriate;

(c) adopt use of sustainable energy sources where feasible;

(d) respond in a timely fashion to changing needs of the organisation;

(e) procure the correct products and service;

(f) motivate and train occupants and energy end users; and

(g) seek third party accreditation for the energy management system.

Business plans and project related business cases are essential tools for any organisations. Typically focussed on financial aspects and logistics, careful business case preparation will justify and explain the financial investment required and outline the expected financial returns.

Adding energy management to these plans should also be considered using similar concepts of energy investment and energy return. This will enable a coherent plan to ensure that any new development:

(a) fits in with the energy policy of the site or installation; and

(b) does not conflict in the overall energy strategy.

Business cases currently focus on financial analysis; energy analysis should also be included.

Large organisations often evaluate the business overheads in terms of cost centres. There is also a similar case model to examine the overall energy budget and the internal energy consumption centres too.

Understanding the load in each area and its effect on the performance of the organisation is very important. Energy use in one energy consumption centre may drive more financial income than the next centre, so the following questions have to be asked:

(a) what is the energy overhead;

(b) is it acceptable given the function of the department; and

(c) what return on that energy investment can be expected?

Energy policy updates

Communication on energy management ensures that:

(a) all staff and visitors are aware of the culture of energy management;

(b) where they can assist with maintenance of the standards that have been set; and

(c) where they can contribute to improvements.

A clear site induction process and regular ongoing training updates are essential to ensure the energy management policy is clear to all who either visit or work on site.

There should be clear feedback processes in place to ensure that everyone in an organisation has:

(a) a clear responsibility to manage energy at some level; and

(b) a clear channel to influence change for the better.

The promotion of an ethos within an organisation that energy management is everyone's responsibility should be considered.

Energy management strategy and policy will be set, signed off and authorised by senior management and other key stakeholders. This needs supporting by regular measuring, monitoring and comparisons to past performance.

A1.2 Energy management strategy

It is important to develop an energy management strategy that is aligned with the energy policy and can be adapted to changing business needs. This includes understanding:

(a) the development of the organisation;

(b) the existing energy supply connections;

(c) the existing energy consumption; and

(d) the future aspirations of the organisation.

An energy management strategy must seek to optimise energy efficiency whilst still allowing some headroom for sustainable business growth. A badly planned or incorrectly executed energy management strategy will possibly inhibit that growth.

	Self-assessment question: What is the energy strategy status?	Yes, or No?
a	The organisation has no energy strategy and reacts to business changes retrospectively. It also is not aligned to legislative requirements.	
b	There is an energy strategy but it lacks resources and active management. It does not always cope with future plans or changes to the organisation's needs.	
c	The energy strategy aligns satisfactorily with the energy policy and legislative requirements. It is reviewed and able to cope in advance of any changes to the organisation.	

Key requirements

The energy manager should understand the key requirements to developing an energy strategy. It is important that any strategy is carefully considered. A number of criteria should be looked at, including:

(a) status of existing energy requirements need to be completely understood;

(b) requirements for future energy demands need to be carefully estimated;

(c) the energy strategy underpins the global energy reduction targets in the energy policy;

(d) use of clear and time bound Energy Performance Indicators (EPIs) to underpin granular targets;

(e) use of a clear procurement policy for energy intakes and also for energy efficient equipment;

(f) checking of existing grid-connected supplies to ensure they are of appropriate capacity;

(g) any technological barriers need to checked (for example, compatibility issues with control systems);

(h) barriers caused by existing working cultures need to be assessed;

(i) working practices need to be examined and training requirements assessed;

(j) options for sustainable supplies need to be understood — either local generation or procurement from further up the supply chain;

(k) any on site requirements for resilience and emergency back-up supplies should be understood — especially where safety of occupants or business continuity can be affected;

(l) any existing legislative requirements that need remedial action;

(m) any future legislative requirements that need planning and resources; and

(n) the roles and responsibilities of key staff who implement and administer the energy management strategy and their associated training requirements.

Understanding priorities

The energy manager should work with key stakeholders in the organisation to establish the main priorities with respect to reducing energy consumption. Consultations should ascertain if:

(a) working practices can be refined;

(b) improved management processes, structures or systems are needed;

(c) staff and visitor awareness or training is needed; and

(d) replacement technology and controls should be deployed.

Priority	Commentary
Financial investment	Energy management solutions can be evaluated against the payback time of the original capital investment. They may also be described as no/low cost, or as having short to medium term or longer term payback. No or low cost usually focus on behaviour change of building occupants and improving work flows to reduce energy use. Short to medium term projects may involve changing some equipment, enhancing local controls or insulating to mitigate energy losses. Longer term controls will involve complete infrastructure upgrades and more efficient designs on the site's energy distribution system.
Financial costs	It is often a requirement that investments in energy saving projects are justified by payback calculations. These may be calculated on either simple payback calculations or using lifecycle cost analysis methods. Such calculations will be necessary for third-party funding channels and it could be advisable to seek external assistance for this. Some project business cases could simply accept that necessary investments need to be made in energy infrastructure. This might: (a) form part of a more general refurbishment programme; (b) be part of a rebranding exercise to demonstrate a more sustainable approach; and (c) be driven by legislative requirements. In such cases payback may be less of an issue. Reduction in carbon use or changing to a more sustainable energy source may be a greater priority to the working ethics of the organisation than simply measuring the payback.

Priority	Commentary
Measurement and verification	In order to set an energy strategy, it is important to understand the metrics by which consumption is measured, the benchmarks by which performance is compared and also the business operational baseline. (a) Where comparisons to previous performances are used, is it possible to verify that the previous and current measurements are realistic? (b) Where more detailed information is required, does the installation have enough energy meters on the right energy systems in the correct locations? (c) Where sub-meters are used for fiscal purposes, do they comply with the appropriate standards? (d) Where sub-meters are used for electrical energy and power quality purposes, are they appropriately specified and calibrated?
Impact on business overheads	It is important to recognise the impact of energy use on business overheads. The energy strategy will need to confirm whether this is determined for the whole business in one go or whether the energy use overhead will be measured and monitored on business cost centre basis.
Impact on business operations	Energy management should monitor energy consumption and seek to improve efficiency. Energy strategies must make clear that successful business operations are aided by efficient distribution of energy and controls that allow the correct amount of energy consumption at the right time.
Impact on business development	Insufficient energy capacity or too tight controls on energy consumption will inhibit business development. Energy strategies must define how to make maximum use of energy whilst controlling excess use and allowing satisfactory levels of business development where feasible.

Understanding barriers

Energy strategies should recognise and address a number of potential barriers to successful energy management. Often though, barriers can be made worse, or better, by the approach of people within the organisation. Energy management needs full commitment at boardroom level. It also needs support and commitment from the management and staff.

Barrier	Commentary
Standards and regulatory requirements	A lack of understanding by both energy management and engineering facilities and maintenance management staff of the relevant standards and regulations that govern the efficient consumption of energy.
	Increasing legislative and regulatory demands are being placed on organisations that need to be recognised and fulfilled.
Political inertia	Within any business there may well be conflicts of priorities, both internally on a departmental level, and externally with any number of other stakeholders.
	Unless resolved there could be any impact on the implementation and successful operation of any energy management system.
	Boardroom champions and properly delegated authority to appropriately qualified personnel are necessary to keep energy management processes on track.
Financial inertia	Financial inertia is the lack of resources to deploy energy efficient technology or experienced personnel to closely manage the consumption of energy.
	Coupled with this could be business cases that prefer capital payback over capital investment when procuring energy saving technologies. This may influence the use of quick wins over longer term solutions.
Skills and knowledge gaps	Skills and knowledge gaps presents a number of barriers: (a) limited education around occupant behaviour; (b) end user familiarity of deployed equipment; (c) inadequate awareness and training around the maintenance of energy saving technology; and (d) a lack of dedicated energy managers to ensure compliance with legislative requirements.
System resilience and redundancy	The requirements for system resilience in safety critical systems will not necessarily assist with an energy efficient installation.
	Some redundancy may be required in the infrastructure and each separate part needs to be efficient.

Barrier	Commentary
Capacity	Restrictions on the existing and potential intake capacity may inhibit future growth. Energy efficiency may assist to make better use of the available resource but there will be limitations at some point.
Legacy equipment	Older installations will inevitably have legacy equipment. Obsolescence and incompatibility with newer equipment may also impede good energy management. This could make close control of energy consumption monitoring difficult and meaningful measuring impossible. There will be a balance between acceptance of equipment that still works versus newer equipment that allows greater controls.
Technical inertia	A similar issue to legacy equipment; technical inertia will see installations attempting to be more efficient but tied to closed protocol systems. Examples may be: (a) early models of building management systems that inhibit wider energy management roll outs due to equipment compatibility issues; and (b) efficiency of existing equipment and their operation such as combustion efficiency of certain kinds of heating systems.
Building ownership (rental scenarios)	The actual occupants of a building or facility do not always have direct responsibility for the energy performance of the site. For example, if an organisation rents offices they may have a single service charge that covers light, heat, telecoms, cleaning and other services. As a result, there may be no incentive to operate the offices efficiently as the occupants would not benefit from any savings. Conversely the landlord's service charge is likely to be more than sufficient to cover the buildings overhead costs, and hence gives no real incentive to make the building more efficient.

Energy hierarchy

An energy management strategy should consider the widely-recognised Energy Hierarchy. Following initial work by Philip Wolfe, the Institution of Mechanical Engineers (IMechE) adapted and adopted the following:

Energy Hierarchy (IMechE 2009)	Headline activity	Commentary for energy managers
Sustainable energy use		
Priority 1: Energy conservation – changing wasteful behaviour to reduce demand	Reduce	Examples include making better use of energy. Tasks include modifying users' behaviour. Outcomes include encouraging an ethos of conserving energy and saves energy at the point of use by changing user behaviour.
Priority 2: Energy efficiency – using technology to reduce demand and eliminate waste.	Replace	Examples include more efficient active equipment and controls or passive measures to improve building fabric and insulation. Tasks include better designs and control philosophies. The outcome saves energy at the point of use by introducing technology.
Priority 3: Exploitation of renewable, sustainable resources	Renew	Examples include using local or off site renewable technologies to supply energy to the site. The outcome may not save energy at the point of use, but energy is derived from a greener source.
Priority 4: Exploitation of non-sustainable resources using low carbon technologies	Revise	Examples include using local CHP or district heating to make reductions in carbon footprints. The outcome may not save energy at the point of use but less energy might be wasted at the source.
Priority 5: Exploitation of conventional resources derived from fossil fuels in the traditional way	Reject	Business as usual – in real terms not actually addressing the energy management problem.
Unsustainable energy use		

Energy loss

Tackling energy loss is an essential component in any energy management strategy. Accepting the procurement of energy just to see heat evaporating through a leaky building or uninsulated mechanical systems should not be tolerated.

An approach often adopted is to tackle building fabric insulation. This can include retrofitting insulation within building voids or additional insulation externally.

Considerable attention should be made to ensure that retro fitted insulation is correctly installed and does not impact on the internal environment causing condensation or poor indoor air quality.

It should also be recognised that pipe and duct work insulation is an effective technology that can give immediate improvements. It provides significant carbon savings and is cost effective. Areas to consider include:

(a) process pipework;
(b) refrigeration pipework;
(c) chilled water pipework;
(d) 'domestic' heating and hot water services;
(e) non-domestic hot water services; and
(f) non-domestic heating services.

It is worth noting that the following references provide advice on pipework insulation:

(a) BS 5422:2009 – Method for specifying thermal insulating materials for pipes, tanks, vessels, ductwork and equipment operating within the temperature range -40 °C to +700 °C.
(b) UK National Engineering Specification – NES Y-50 (2011).

Likewise taking no action to reduce inefficiency that has been identified in electrical systems is also not acceptable.

Careful analysis of energy losses is advisable, especially when business cases are also necessary to allow for financial investment in reducing those energy losses. Some activities may provide for greater savings than others. Finance departments will often want to see that investment in energy waste reduction can provide a return on that investment after a certain period of time.

Reduce energy losses

As well as the education of end users, both staff and visitors alike (see Appendix A2.3: User behaviour), the reduction of energy waste may involve introducing passive technology improvements. This could be improvements to the insulation of the building fabric. The benefits of such an approach inside a building will include:

(a) reduction of heat loss in cool external conditions and hence less demand on the heating system; and
(b) reduction of heat gain in warmer external conditions and hence less demand on the air conditioning for comfort cooling.

At a system level, insulation of space heating pipework or ductwork, insulation of water heating pipework and cladding of hot water storage tanks assists in keeping energy where it is needed instead of leaking in an uncontrolled fashion.

Ideally these will be designed into new installations and industry standards will enforce the implementation of such measures. In older installations, retrospective fitting of insulation should be considered.

> **Caution:**
> Professional engineering advice is advisable to ensure the deployment of passive technologies that will actually help and not cause additional problems – simply making a building airtight without understanding the implications of adequate ventilation and the requirements for fresh air will produce an unhealthy environment for the occupants. Correct deployment of passive technologies will assist energy management strategies.

Reduce inefficiency

An engineering solution is the introduction of active technology improvements to engineering systems to reduce system losses. This will include:

(a) power factor correction to electrical switchboards and principal loads;

(b) harmonic filters to motor control panels that are connected to variable speed drives;

(c) heat exchangers to water systems;

(d) heat recovery to ventilation systems;

(e) occupancy controls; and

(f) improvements to infrastructure efficiency by use of barycentres and better distribution.

> Caution:
> Each aspect will need careful design. Some examples are:
> **(a)** over-specifying power factor correction, without active controls, will not necessarily provide the necessary improvements compared to the capital cost involved; and
> **(b)** providing variable speed drives within a plant room may improve the performance of each individual motor, but the impact of the harmonic currents on the building infrastructure could be far worse.
> Solving one problem in isolation is not necessarily helpful and could cause other problems in other parts of the infrastructure.
> It should be noted, though, that correctly engineered design solutions will improve energy efficiency.

The impact of energy waste reduction

Figure A1.1 below, demonstrates the impact that reducing waste by passive measures and also through more active response can have on the energy equation. As waste reduces, the energy input that previously fed that waste decreases in equal proportion.

Reduce waste energy — introduction of passive technology could include:

(a) improvements to the building fabric; and

(b) insulating pipework and ductwork and insulation to hot water tanks, to reduce:

    i loss of heat in cool external conditions; and

    ii reduce requirement for cooling from solar gain in warmer external conditions.

Reduce inefficiency — introduction of active technology improvements to engineering systems to reduce system losses, could include:

(a) power factor correction to electrical switchboards and principal loads;

(b) harmonic filters to motor control panels; and

(c) heat exchangers to water systems.

▼ **Figure A1.1** Reduce energy waste and energy inefficiency

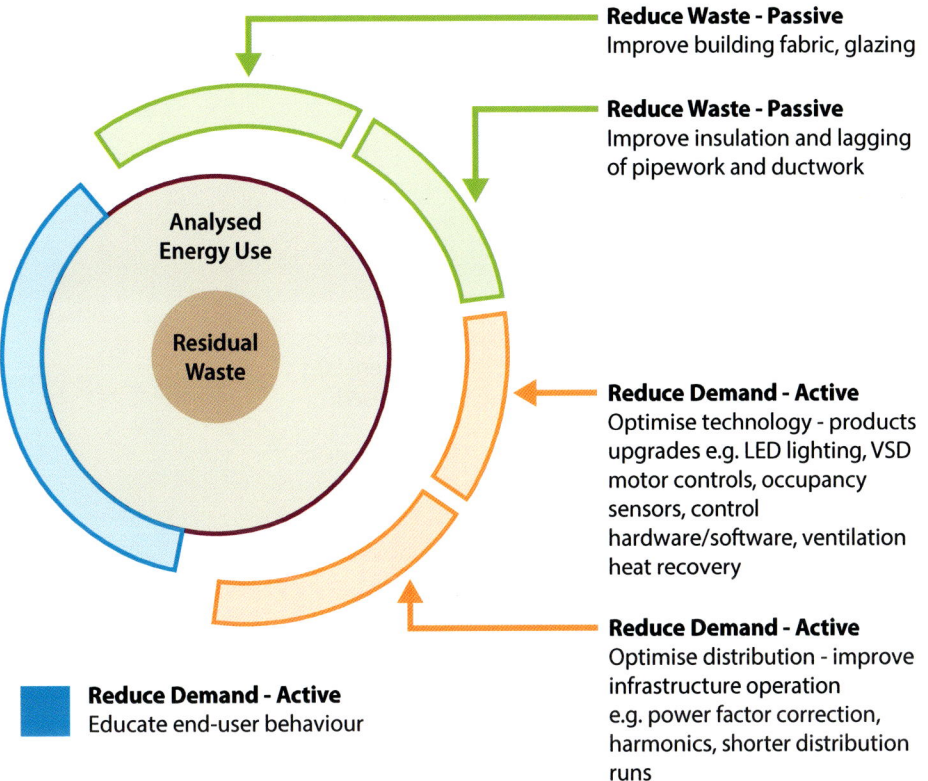

Reduce Waste - Passive
Improve building fabric, glazing

Reduce Waste - Passive
Improve insulation and lagging
of pipework and ductwork

Reduce Demand - Active
Optimise technology - products
upgrades e.g. LED lighting, VSD
motor controls, occupancy
sensors, control
hardware/software, ventilation
heat recovery

Reduce Demand - Active
Optimise distribution - improve
infrastructure operation
e.g. power factor correction,
harmonics, shorter distribution
runs

Reduce Demand - Active
Educate end-user behaviour

Making practical use of greater efficiency

Reduction of waste and inefficiency may be to allow the net user demand to actually increase without raising the site supply capacity.

With electrical supplies, reducing reactive energy losses (kVAr) will allow an increase in real power (kW) without necessarily increasing apparent power (kVA) at the meter of the existing input. This could be useful where there are local supply issues when demand does increase. Such a strategy allows an installation to develop to a certain degree without costly infrastructure improvement projects.

The risk with this approach is that real demand is not analysed to ensure this is also optimised. Aspects such as user behaviour, optimising controls, devices and energy recovery will not necessarily be dealt with in this manner.

▼ **Figure A1.2** Utilising recovered capacity

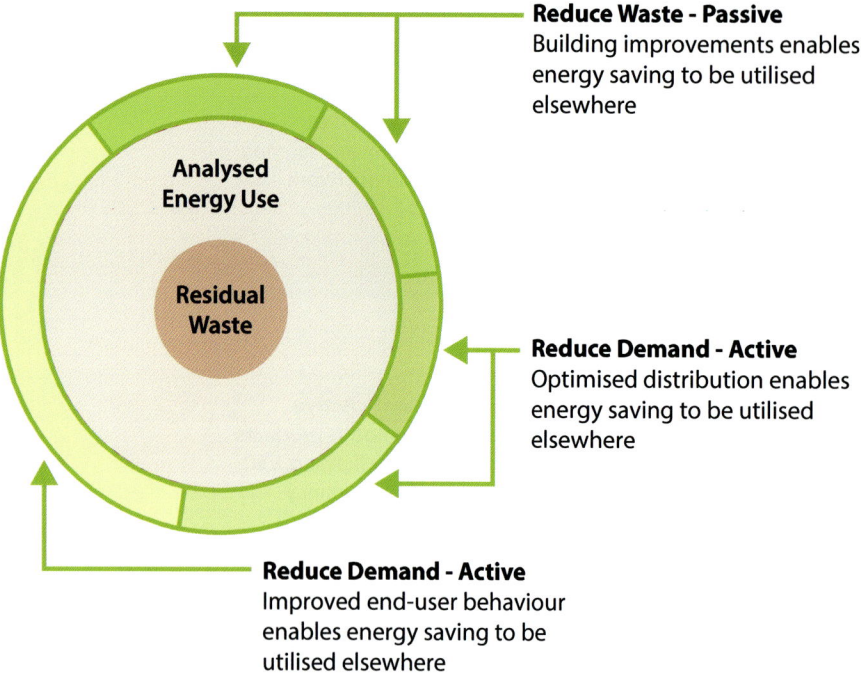

Reduce Waste - Passive
Building improvements enables energy saving to be utilised elsewhere

Analysed Energy Use

Residual Waste

Reduce Demand - Active
Optimised distribution enables energy saving to be utilised elsewhere

Reduce Demand - Active
Improved end-user behaviour enables energy saving to be utilised elsewhere

Reducing demand and inefficiencies can assist in allowing more load to be connected to the site infrastructure without necessarily increasing the overall site capacity

Drawbacks of reducing demand

By reducing waste in use and inefficiency in distribution, carefully controlled energy management systems can potentially reduce energy consumption and hence carbon emissions. However, the savings might not be large enough to convince the organisation's financial team to invest.

Proper consideration of a wider strategy should be made. If solutions are provided in isolation, there is there is a risk that energy consumption could actually increase overall.

There is a need to manage energy today; there is also a need to radically change how energy is used in the future at all ends of the supply chain.

Caution:

The following are examples of fundamental problems with approaches that do not manage energy in a holistic way:

(a) simply swapping lighting from older technology to LED — this does not change the way people use artificial illumination or switch off when not in use;

(b) making use of newer technology, for example, LED lighting — such an approach makes use of smaller forms and designs but this can result in a proliferation of additional lighting being used and consequently have the opposite outcome to lowering carbon emissions;

(c) the central corporate philosophy at board level may still revolve around the assumption that the existing energy landscape will continue as it has always done, fossil fuels will continue to provide and society just needs to be more efficient with it;

(d) the central corporate philosophy may take the view that a decarbonised grid is a national problem and being resolved by others, all they need to do is pay for a green tariff — this does not account for the efficiencies in use that are still required by the business; and

(e) continued use of fossil fuel based energy — this approach does not prepare society for a world where further extraction of this energy source is no longer financially viable or ecologically desirable.

Using electric cars may salve the conscience of the motorist, and provide a new revenue stream for motor companies, but the electricity used still needs to be generated somewhere. Careful consideration will ensure that this is not a problem simply being passed to another part of the supply chain.

Improving technology in electric cars should result in more efficient use of energy for travelling, coupled with more efficient and ecologically sound methods of electrical energy generation and transmission.

Energy management strategies that focus on behaviour, control and technologies in a coherent manner will save energy and improve user experiences.

A1.3 Energy management procedures

Identifying and controlling your energy risks

What is an energy management system in practical terms?

What is a good practice energy management lifecycle process?

What are the risks to those ideas?

The concepts and analysis of operational risk management are well established in the health and safety field of industry. Design risks are also analysed carefully and mitigations provided. Similar philosophies should be in place for energy management.

Within the built environment design tools such as BREEAM and LEED allows risks specific to energy consumption and energy management to be assessed and mitigated.

	Self-assessment question: What is the energy procedures status?	Yes, or No?
a	The organisation has no energy procedures of its own and does not use recognised standards or industry codes to manage energy at an operational level. It also is not aligned to legislative requirements.	
b	There are some energy procedures but they are not complete or lack resources and active management. The procedures do not match the organisation's needs.	
c	The energy procedures align satisfactorily with the energy policy, energy strategy and legislative requirements. They are reviewed and able to adapt to any changes to the organisation.	

Health and safety model

The Health and Safety at Work Act (HASAWA) puts varying degrees of responsibility on everyone to ensure health and safety in the workplace. Within the safety hierarchy, different stakeholders have specific responsibilities in an organisation to ensure the safety of employees, visitors and the general public.

Safe systems of work, based around operational risk assessments and controlled by signed-off safety permits, provide a framework that assess risks and provide mitigation to enhance safety. On building sites and in other similar environments, health and safety is constantly updated by the use of 'tool box' talks, where the latest information and metrics on safety are given.

These safety engineering models should be adapted for energy management and then adopted to improve decision making. Risk analysis, directly related to energy management, will ensure greater efficiency and increase awareness of energy conservation.

Energy safety model

Loss of energy through wasteful practices and infrastructure should be seen as a risk to the environment, to the operation of the installation and to the finances of the organisation.

A business operating model, focusing on energy management procedures, should be set in place by the senior levels of an organisation, underpinned by board level leadership and properly communicated to the maintenance teams, engineers and other operative staff.

Risk assessment pictorial analysis

For operational decisions on energy management, ISO 50001 is a recognised starting point. Taking each area in turn, the topics can be analysed for their respective impacts on energy management and assigned a score.

It is probable that existing sites with legacy installations, on initial assessment, will fall within the red part of the circle shown below. As continuous improvements are made to processes and reported in each area of the organisation and installation, the colour changes towards green.

Newer sites, probably designed and built to adhere to BREEAM, for example, would be expected to score higher at initial assessment as better designs, newer equipment, robust maintenance processes and associated documentation should be already in place at handover.

48

The pictorial analysis below, based on the principal headings of ISO 50001, shows how different parts of the process can be addressed separately whilst being an integral part of a much wider picture. Through a process of measurement, planning and continuous improvement they are each being drawn towards the centre. This can be used to demonstrate that the energy management process is improving the efficiency of the installation and hence the system is on target.

▼ **Figure A1.3** Targeting improvements with BS EN ISO 50001

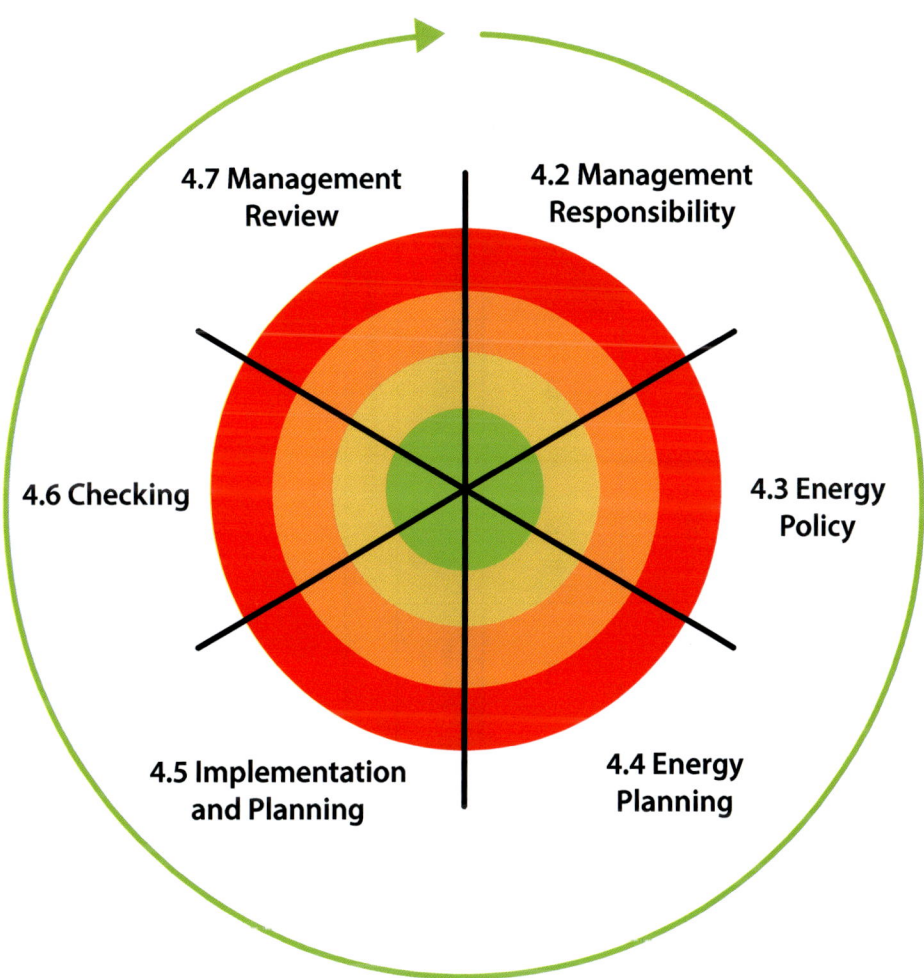

Numerical analysis

The energy risk analysis table below provides an analysis and reporting mechanism to help demonstrate what stage of development that element of the overall energy management plan has reached.

As each stage is developed, improvements will be achieved. These in turn reduce the energy consumed and the tariff score can be reduced too.

Operational level					Implementation level
No progress	Not operating satisfactorily	Improvements acknowledged	Improvements in place	Operating satisfactorily	
5	4	3	2	1	Plan operating & continuous monitoring
10	8	6	4	2	Plan in place ahead of initial review
15	12	9	6	3	Plan identified with further assessments
20	15	12	8	4	Plan identified with initial appraisal
25	20	15	10	5	Plan not in place

Criteria	Explanation
Operating satisfactorily	An energy management plan is in place and has been operating satisfactorily for several regular periods. A continuous pattern of reporting exists and has been analysed for further improvements. A robust audit trail exists.
Improvements in place	An energy management plan is in place and has been operating for several regular periods. A continuous pattern of reporting exists and has been analysed. Plans for further improvements are complete. A robust audit trail exists.
Improvements acknowledged	An energy management plan is in place and has been initiated. A pattern of reporting exists and has been analysed. Plans for further improvements are being developed. An audit trail exists.
Not operating satisfactorily	An energy management plan is in place, but has just been initiated. A pattern of reporting does not yet exist so it has not been analysed. Plans for further improvements have been outlined but not developed. No audit trail exists.
No progress	No energy management plan is in place. No energy policy or strategy has been identified. Plans for further improvements are at the concept stage. No audit trail exists.

Criteria	Explanation
Plan operating with continuous monitoring	An energy management plan is in place and has been operating satisfactorily for several regular periods. A continuous pattern of monitoring exists. Energy management activities have been fully embraced by the organisation.
Plan implemented ahead of initial review	An energy management plan is in place and has been operating for several regular periods. A methodology for monitoring exists and is being implemented. Outline plans for further improvements are being developed further. Energy management activities are accepted by the organisation.
Plan identified with further assessments	An energy management plan is in place and has been initiated. A methodology for monitoring exists and awaits full implementation. Plans for further improvements need to be developed. Energy management activities are being promoted to the organisation.
Plan identified with initial appraisal	An energy management plan is in place, but has just been initiated. A methodology for monitoring has been agreed and awaits initial implementation. It is too soon for plans for further improvements to be developed. Energy management activities have yet to be promoted to the organisation.
Plan not in place	No energy management plan is in place. Monitoring is restricted to main utility energy meters. There is no ownership of plans for further improvements. There is no active promotion of energy management to the organisation.

A1.4 Engineering design for energy management

All installations begin as design drawings. Embedding energy management concepts firmly within the design is necessary to ensure that energy is correctly consumed from the outset. The energy manager must understand that the performance of different components within an installation are governed and guided by various documents.

Retrospective refurbishments and alterations are often influenced by pay back calculations and governed by similar financial criteria.

Self-assessment question: What is the interface to engineering design? Has energy management been considered for new systems, new builds, or retro fit projects?		Yes, or No?
a	There is no direct involvement by the organisation's energy management team as stakeholders in the design of new engineering systems.	
b	There is some consultation of the organisation's energy management team for new designs, but it is inconsistent. The process does not always match the organisation's needs or the energy management team's expectations.	
c	The engineering design process aligns satisfactorily with the energy policy, energy strategy and legislative requirements. It is reviewed and projects take into account the existing infrastructure and perceived energy risk assessments.	

In order to ensure continuous sustainable operation and minimise energy consumption three principal factors should be addressed correctly at the design stage:

(a) buildability – designs should carefully consider how they will minimise energy consumption through careful selection of suitable energy efficient products, control philosophy, and construction techniques;
(b) maintainability – designs should consider how the installation will be maintained for ease of access and to ensure that sustainable operation continues in line with the original design concepts; and
(c) disposability – designs should consider how products will be disposed of. Within a building, components may need replacing because they have failed or there is a change of use.

▼ **Figure A1.4** Design factors for energy management systems

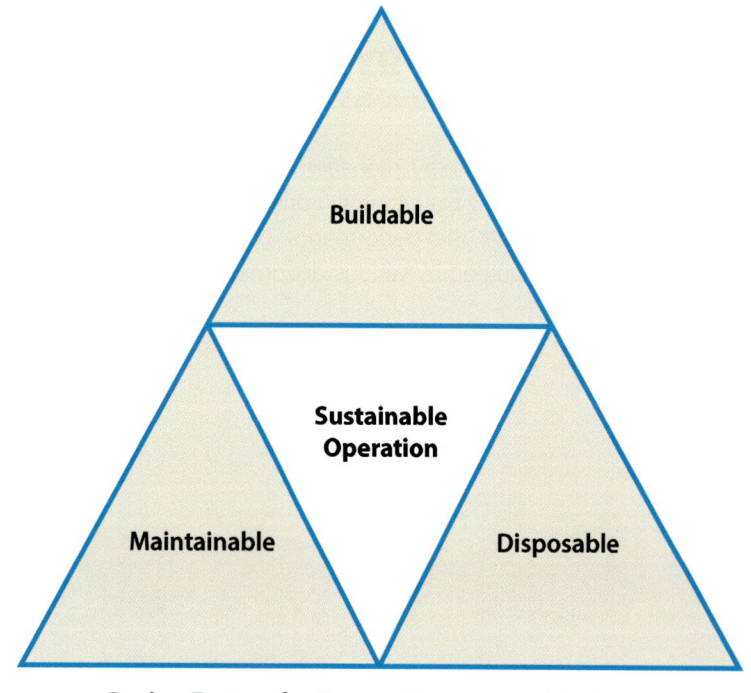

Design Factors for Energy Management Systems

Operational energy management

When the installation is completed satisfactorily to the required environmental design standards, the real challenges for energy management commence. The commissioning team need to ensure that:

(a) the installation works as the design intended; and

(b) the maintenance team and end users are properly briefed on their roles in making sure that the installation continues to operate in the manner intended.

Operations and maintenance manuals should provide full details of the installation to enable a comprehensive programme of activities to be compiled. These regular maintenance tasks will provide reassurance that the installation is working, either as intended, or providing feedback to address remedial actions.

The design stage should have adequately considered how energy consumption will be measured, the data collated and the manner it will be presented to the primary stakeholders. These will be the site based personnel subsequently charged with operational responsibility and who will be paying the energy costs.

Disposability and energy management

Energy management systems need to consider how replaced products will be disposed of. Maintenance teams need to consider this aspect from safety and environmental considerations too.

Using design risk assessments, and some experienced foresight, it should be feasible to reduce energy consumption when replacing and disposing of old products.

It is important to also remember that there are the environmental considerations of BS EN 14001 in this respect. Health and safety considerations also need addressing in terms of the removal of redundant equipment if the items are heavy or installed at high level.

Product design and technology advances at a tremendous rate. However, it is important that products at the end of their particular useful life can be recycled to recover rare materials or upcycled to use as part of another product.

Installation designs should focus on the use of products that can be easily replaced so that the system can continue to function.

The failure of one component should not mean the replacement of complete infrastructures because of 'backward compatibility issues'. Such a conflict will impact on wider energy management.

Low or no maintenance products often mean that the expected life cycle of the product is not great and the item is disposable. The economic and energy value of such products needs to be carefully considered.

The following case study first appeared in BRE Information Paper IP7/13 and is reproduced by kind permission of the original Author Dr Andy Lewry of BRE.

The BRE information paper in general, and the case study in particular, highlights the need for proactive leadership of the company hierarchy in understanding the purpose of energy management and in driving the agenda forward.

Camfil UK is the global leader in providing air filtration and clean air solutions. The company's filters are manufactured at 23 production plants across the world and are used in offices, mines, factories, hospitals, nuclear power stations and more. The company is run from a 160,000 sq. ft. manufacturing plant in Lancashire with approximately 200 employees – and the entire workforce is committed to managing and saving energy. The company leads by example and became the first UK manufacturer to achieve BS EN 16001:2009[20] and in 2011 transitioned to ISO 50001:2011[2].

Camfil wanted to be seen to be practising what it promotes, and the company established basic management systems for reporting monthly energy consumption across gas, electricity and logistics – using air leak detection surveys and thermal imaging cameras, as well as an online toolkit to monitor progress and calculate savings made.

Getting started

Camfil was fortunate in that it already had an energy audit procedure in place. Quality, Environmental and Energy Manager Brian Haslam explained: "The most challenging part of ISO 50001 is the implementation and ongoing monitoring of energy use against other relevant variables. This involves a long-term commitment to monitoring and allocation of resources." This proved beneficial as the company began to look at its energy use in more detail and identified previously unmonitored areas of energy use. This in turn allowed Camfil to achieve savings that might otherwise have been overlooked. Brian added: "We believe if you can't measure your energy, you can't save it."

Energy saving

The Camfil Energy Awareness Saves Environment (CEASE) campaign achieved significant reductions in energy usage and improvements in energy efficiency, resulting in the company saving some £200K+ on energy bills through low cost, self-funding opportunities.

Two main categories of energy usage were identified: logistics fuel and electricity.

The savings that have been achieved are summarised as follows:

	2008	2009	2010	2011	2012
Logistics Fuel		-23 %	-7 %	-1 %	-10 %
Electricity	-14 %	-8 %	-8 %	-6 %	-4 %

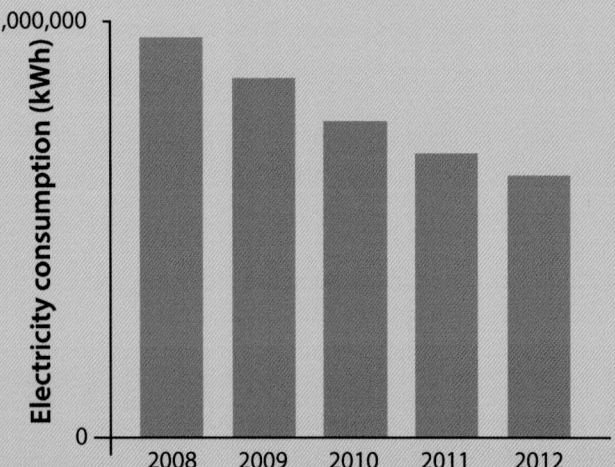

The overall result is that Camfil has reduced its energy costs from £500,000 in 2008 to £300,000 in 2012.

This case study was reproduced by kind permission of Camfil Farr UK Ltd.

Managing procurement, resources and people

A2.1 Procurement and energy

Energy costs are affected in a number of ways. The energy manager should use their influence on the organisation to ensure correct procurement of all stages from energy intake at the appropriate tariffs, to energy efficient distribution systems, to energy efficient devices and controls connected at the point of use.

Procurement, though, is only one aspect of the role of the energy manager. Simply accepting the suppliers' energy bills at face value and processing payments is not enough.

Further advice in this area is available from:

1   Energy Managers Association — *Energy Managers' Guide to Electricity Procurement;* and
2   Carbon Trust (CTG054) — *Energy Management.*

Other considerations are:

1   Ecolabel;
2   Green Public Procurement;
3   ErP (Ecodesign of Energy related products); and
4   Energy Label.

Evaluation of your current business status — energy supplies

It is important for an energy manager to have a procurement plan with respect to energy supplies.

	Self-assessment question: What is the energy procurement plan?	Yes, or No?
a	Energy supplies are purchased directly from the energy suppliers and user equipment is purchased as and when required.	
b	There is some consultation of the organisation's energy management team but it is inconsistent. The process does not always match the organisation's needs or the energy management team's expectations.	
c	The energy procurement process aligns satisfactorily with the energy policy and energy strategy requirements. It optimises the available tariffs and the use of energy efficient equipment.	

There will be a number of key factors for the energy manager to understand, as well as the implications of those factors on the running of the organisation. Energy costs ultimately can undermine business profitability.

Procurement factor	Commentary
Capacity	Does the connected grid supply capacity match the overall demand? Too little and there will be a lengthy procurement process to secure a larger supply or a second supply to the same location. Too large and the organisation will be paying a premium for a standing charge on the larger capacity that it is not using.
Fixed or flexible	Energy managers should understand the types of energy contract on offer. Depending on the procurement route energy costs can be on a fixed or flexible contract. Energy costs can fluctuate and whilst the trend is typically upwards, over any given period it may reduce periodically. Appetites to risk and cost certainty will dictate whether to go flexible or fixed. Impartial professional advice should be sought.
Market route	There are a number of options for an organisation to use when procuring energy. Choices will be made depending on the size of organisation and also influenced by the energy policy and/or strategy of the business. (a) Dealing direct with suppliers. Some have dual fuel supplies and a selection of tariffs and contract lengths to suit. (b) Dealing through an energy broker. These will seek out energy deals on behalf of the organisation and may be able to take advantage of aggregating multiple clients' demands to get improved terms. (c) Dealing through a Central Purchasing Body. Typically used by government departments to aggregate multiple organisation demands to improve the terms of all. Energy managers are key stakeholders in any negotiations with respect to energy contracts and for new supplies.
Overheads	Energy costs will have an impact on business overheads. Sub-metering will enable cost centres within the organisation to assess their own use and manage energy consumption locally. For multiple site operations it will be important to have local information for analysis and improvements.
Billing	Energy managers should be able to interpret the information on the energy bills. There should be a clear understanding of which parts of the bills can be improved (based on kWh) and which cannot be changed (based on standing charges).
Tariffs	For electricity supplies in particular, tariffs can vary according to the contract agreed and the time of day that energy is consumed. The energy manager should understand what these timings and tariffs are – costs during peak and off-peak hours. They should also have analysed what electrical loads are on the lower tariff and whether other loads can be moved to suit.

Procurement factor	Commentary
Engineering design	As discussed in Appendix A1.4, engineering design and infrastructure efficiency will impact on energy consumption and hence costs. Energy managers are key stakeholders in this element of a building's services procurement. They should be satisfied that efficient distribution, adequate controls and low energy devices are included.
Equipment efficiency	Energy managers should influence and monitor the procurement of energy efficient fixed equipment, associated controls and portable appliances to reduce energy consumption, and hence costs. New legislative documents such as the European Energy-related Products Directive encourage the introduction of newer, more energy efficient devices within an engineering system to reduce maximum demands. Examples include LED lighting, modern radiators and other heat emitters, energy efficient boilers, hybrid air conditioning systems and heat recovery systems.
Water management	Notwithstanding the inevitable costs of water use and sewage, within large organisations, water consumption will also be an important part of process and production. Uncontrolled water use and large system losses can also have an indirect influence on energy management issues too where pumps are running to keep up with demand.
Maintenance	Although not part of the energy manager's direct remit, timely maintenance will influence improvements in energy consumption. Coordination with the maintenance team to ensure timely work on air conditioning ahead of the summer season, or boilers ahead of the winter season, will help to reduce losses. Procurement of appropriately trained resources, either in-house or external contractors, should involve the energy manager as a key stakeholder. Careful selection of energy efficient replacement parts, which may not necessarily be the lowest initial procurement cost, should assist in reducing ongoing energy costs and maintenance. It should also assist with whole life cycle costing.
Seasonal use of energy	Energy managers should be aware of seasonal trends in energy consumption and where appropriate adjust supplies accordingly, for example where water heating and space heating is driven by oil delivered to site.
Market trends	Market trends in prices for oil fuels will also influence procurement costs, where local tank storage is used.

Procurement factor	Commentary
Resilience and back-up supplies	Energy managers should be aware of any services infrastructure that may need additional resilience and plan accordingly. Some installations, for example, will have the need for standby generators or backup supplies to be installed. The impact if these fail to start may be severe on an organisation's ability to function in the event of interruptions to grid supplies. Energy managers should liaise with maintenance staff to ensure that such equipment is subjected to careful monitoring and regular testing. This will include: (a) the mechanical and electrical aspects of a generator; and (b) the shelf life and risks of contamination of locally stored fuels.

Understanding gas bills

Gas meters measure volume of gas drawn from the grid-connected network. Additional factors are then required to convert a reading from a gas meter read into kilowatt hours which is then chargeable.

The calculations used to generate gas bills are demonstrated in the Gas (Calculation of Thermal Energy) Regulations (SI 1996/439). The Office of Gas and Electricity Markets (Ofgem) have responsibility for these regulations.

Cost of energy =

$$\frac{\text{(volume of gas used)} \times \text{(calorific value)} \times \text{(temperature and pressure factor)}}{\text{kWh factor}} \times \text{unit price}$$

Bill component	Commentary
Volume	Gas meters measure volume in cubic feet (imperial measure) or cubic metres (metric measure). Bills require cubic metres for their calculations. If the meter is in cubic feet then the resultant energy use (subtraction of the current reading from the previous reading) is multiplied by either of: (a) 1 cubic foot = 0.0283 cubic metres; or (b) 100 cubic feet = 2.83 cubic metres.
Calorific value	Calorific values can vary during each billing period and are determined nationally. The figure is quoted in megajoules per cubic metre (MJ/m^3). Gas transporters are required by the regulations to maintain a range of 38 MJ/m^3 to 41 MJ/m^3 Figures outside of this range may cause problems with gas appliances.
Temperature and pressure	A conversion factor of 1.02264 is used to compensate for changes to the volume of gas caused by the effects of temperature and pressure in the distribution pipework. The conversion factor is detailed in the regulations.
kWh factor	A constant of 3.6 is used to convert the values to kilowatt-hours for billing purposes.
Unit prices	This will be the agreed cost per unit (kWh) according to the contracted tariff.

Other charges on the bill will include:

(a) Standing charge — a fixed charge typically in pence per day. Where included this charge will be the same in each billing period irrespective of the level of consumption in that period. Standing charge accounts for the cost of metering and the fixed charges of Gas Distribution Networks and National Grid.

(b) Climate Change Levy (CCL) — this is a tax on the energy use of all business customers, and is based on kWh quantity used in the period. CCL aims to promote energy efficiency and reduce greenhouse gas emissions. Current CCL rates can be found on the HMRC website.

(c) Value Added Tax (VAT) — Government tax based on the value of the supply of goods and services. Businesses pay VAT at 20 % on gas supplies. Domestic properties pay VAT at 5 %. Businesses using less than 145 kWh per day pay VAT at 5 %.

Further details on UK government taxes on fuel can be found on the Gov.uk website and in particular using the search term: VAT Notice 701/19: fuel and power.

▼ **Figure A2.1** Basic gas bill based on fixed prices

A.N. Office Block
Main Street
New Town
Anyshire
AB12 3CD

	Invoice period–start	1st May 2016
	Invoice period–finish	31st May 2016
Tax point date	6th June 2016	
Meter number	D76E54321	

Explaining your gas statement	
Formula: Conversion of gas units to kWh using the following formula:	
Meter Units: Value of the current reading less the previous reading	970
Volume conversion factor: To convert meter units to metric. For imperial meters = 2.83, For metric meter = 1.00 Resulting metric units	1.00 970
Volume correction: Gas regulations compensation for changes in gas volume based upon temparture and pressure. Industry standard correct factor is 1.02264.	1.02264
Calorific value: Measurement of the energy content of gas varies throughout the year.	39.4
Convert to kWh: Multiply metric units by volume correction and calorific value and then divide the resultant value by 3.6 to find the number of kilowatt-hours.	3.6
kWh	10,856.45

Item	Units	Rate (p)	Amount (£)
Day units (based on kWh)	10,856.45	2.052	222.77
Standing charge (monthly)	1	233.54	233.54
CCL – climate change key (based on kWh)	10,856.45	0.1950	21.17
Sub total			**477.48**
VAT		20%	95.49
Total			**572.97**

In Figure A2.1, all distribution and transmission charges are inclusive of the rates used.

Understanding electricity bills

Electricity meters for smaller installations measure energy consumption in kilowatt hours (kWh). The bill is then calculated by multiplying use by a unit price. The final cost has standing charges, usually a daily rate, added to it.

For larger installations measurement of electricity, and hence cost, is complicated with additional factors such as maximum demands, transmission charges and distribution charges. Penalties may also be applied for installations with poor load characteristics.

The calculations used to generate electricity bills will vary depending on the particular tariffs the organisation has signed up to. For fixed cost tariffs add-ons such as transmission, distribution and third party costs are hidden. For tariffs, with pass through costs, other items will appear — of these some will be based on the amount of energy consumed.

Bill component	Commentary
Distribution network operator (DNO)	The electricity supply company that provides the local electrical distribution infrastructure network at high or low voltage. This is a distinct entity form the meter suppliers and hence electricity bills, although they could be part of the same parent company.
Distribution use of system (DUoS)	DUoS rates are charged by the DNOs and set against their part of the operation, maintenance and development of the UK's electricity distribution networks. The charges comprise of various elements, including: (a) available capacity; (b) standing charge; and (c) units (typically split into red, amber and green). The DUoS charges may be visible on the electricity bill, or part of the contracted rates.
Transmission network use of system (TNUoS)	TNUoS rates are charged by the transmission network operators and set against their part of the operation, maintenance and development of the UK's electricity distribution networks. Half hourly customers are billed against the TNUoS on a system of triads — matching the maximum demand on their sites on three particular half hours when the demand is at its highest.
Available capacity	This confirms the maximum capacity (kVA) of the site and a holding charge is levied against reserving that capacity from the supplier. If the available capacity is too large and the organisation has no plans to expand the site, then the availability charge could be a disproportionally large overhead on the bill.

Bill component	Commentary
Meter point administration numbers (MPAN)	MPAN are unique numbers for each and every meter. |	Profile Class	Meter Time Switch Code	Line Loss Factor
S	01	234	567
	10	2345	6789	012 | Distribution ID — Meter Point ID number — Check Digit
Profile class	This two-digit code, part of the MPAN, shows what kind of meter is installed. Domestic customers use classes 01 and 02. Non half-hourly metered sites currently use 03, 04, 05, 06, 07, and 08. Half-hourly metering for large consumers use profile class 00.
Transition to P272	Energy managers should be aware that electricity meters in profile classes 05 – 08 must supply energy usage data on a half-hourly basis. Premises with a half-hourly meter already installed will have been charged half-hourly since November 2015. Remaining premises should be changed and registered by April 2017. Affected businesses must change their metering processes and appoint an accredited data collection (DC) agent and meter operator (MOP).
Half-hourly metering (HH)	Industrial and commercial sites with larger energy requirements will be provided with half-hourly meters that provide electricity reading every half an hour.
Non half-hourly metering (NHH)	A simpler electrical energy meter, usually analogue, that provides information primarily based on kWh. Some older analogue meters may also provide information on maximum demands (kW) during the billing period.
Unit prices	This will be the agreed cost per unit (kWh) according to the contracted tariff.
Peak and off-peak	Smaller organisations will have dual tariffs. Good use of off-peak tariffs for appropriate electrical loads will reduce running costs. Efficiency, design and adequate controls are still required even though running costs are proportionally lower.
Renewables obligation	Supports commercial scale renewable energy electrical energy projects in the UK.
Feed in tariff	Supports government programmes promoting a range of small-scale renewable and low carbon electrical energy technologies.
Contracts for difference	Supports government initiatives to encourage investment in low carbon generation. Investors receive a guaranteed income for the electricity that is generated.
Capacity market	Supports organisations that invest and agree to generate electricity to ensure there is enough capacity at peak demand. Also aimed at large users that are able to reduce electricity consumption when demand is high on the grid.

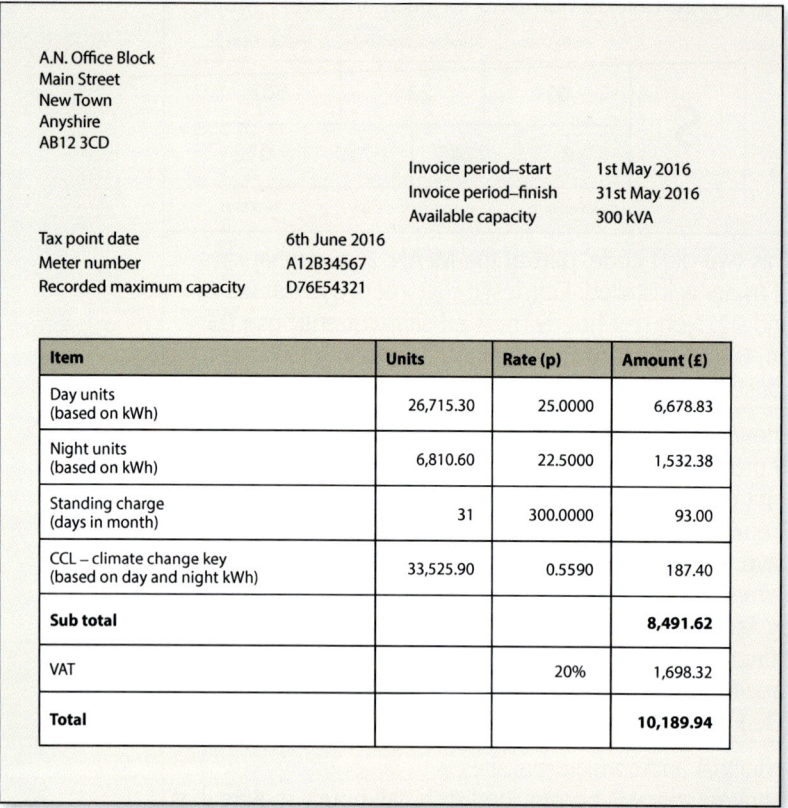

A.N. Office Block
Main Street
New Town
Anyshire
AB12 3CD

Invoice period–start	1st May 2016	
Invoice period–finish	31st May 2016	
Available capacity	300 kVA	

Tax point date	6th June 2016	
Meter number	A12B34567	
Recorded maximum capacity	D76E54321	

Item	Units	Rate (p)	Amount (£)
Day units (based on kWh)	26,715.30	25.0000	6,678.83
Night units (based on kWh)	6,810.60	22.5000	1,532.38
Standing charge (days in month)	31	300.0000	93.00
CCL – climate change key (based on day and night kWh)	33,525.90	0.5590	187.40
Sub total			**8,491.62**
VAT		20%	1,698.32
Total			**10,189.94**

In Figure A2.2, all distribution and transmission charges are inclusive of the rates used.

▼ **Figure A2.3** Simple analysis of half hourly graphs

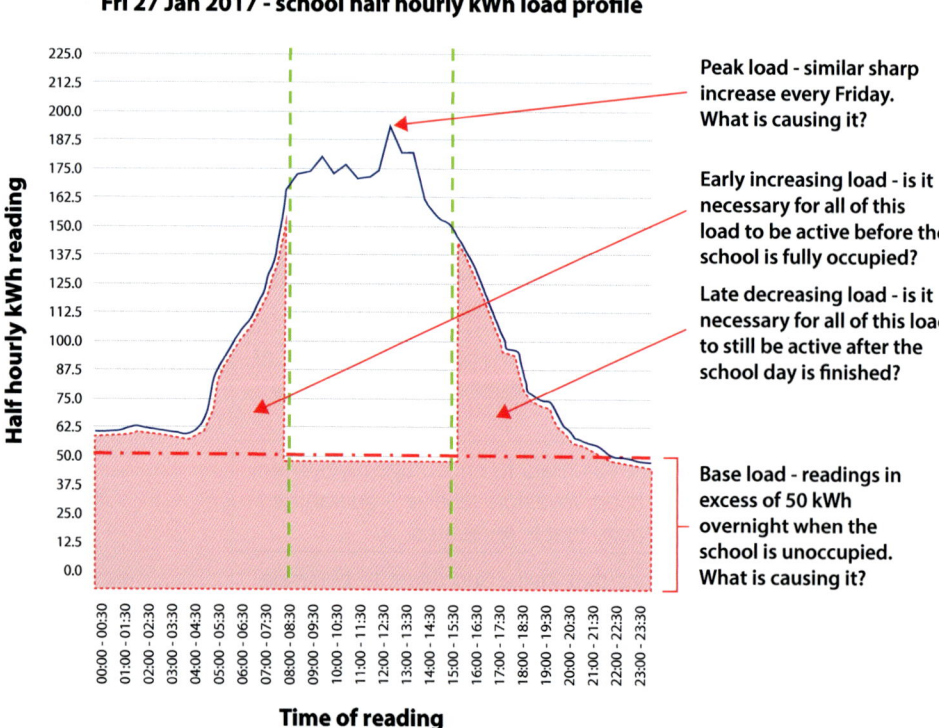

Fri 27 Jan 2017 - school half hourly kWh load profile

Peak load - similar sharp increase every Friday. What is causing it?

Early increasing load - is it necessary for all of this load to be active before the school is fully occupied?

Late decreasing load - is it necessary for all of this load to still be active after the school day is finished?

Base load - readings in excess of 50 kWh overnight when the school is unoccupied. What is causing it?

The graph above (Figure A2.3) shows a simple analysis of half hourly data and associated graphs from a school in the south of England in January 2017.

Such an approach can enable targeted questions to be raised to inform the overall energy management system. The school is open in the day only and not boarding. Targeted improvements can be made, for instance:

(a) investigate why so much load is being used overnight;

(b) investigate why so much load is switched on before the school day; and

(c) investigate why so much load is left on after the school day.

Example questions:

(a) does user behaviour need improving to switch off at the end of the working day;

(b) do the cleaners procedures need to include lighting off;

(c) are the overnight timing controls wrong in remote buildings, especially with electric only heating;

(d) how long does it take to heat the school to satisfactory levels in the morning – can the lead in time be reduced;

(e) what after school hours activities require so much energy use – can this be optimised and/or manually overridden; and

(f) is there a different control strategy for term time and holiday periods?

It will be necessary to adapt investigations to suit local needs but Figure A2.4 demonstrates:

(a) what improvements may be achieved;

(b) what effect these improvements could have on the half hourly graph; and

(c) the overall reduction in energy consumption that could be achieved.

▼ **Figure A2.4** Targeted improvements using analysis of half hourly graphs

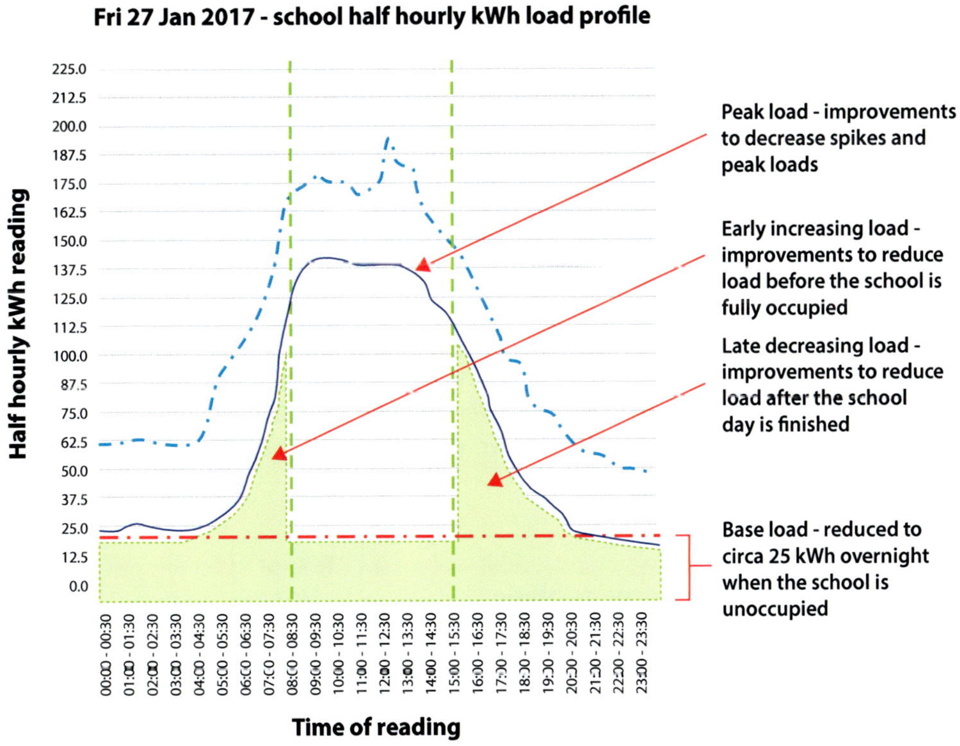

Fri 27 Jan 2017 - school half hourly kWh load profile

Peak load - improvements to decrease spikes and peak loads

Early increasing load - improvements to reduce load before the school is fully occupied

Late decreasing load - improvements to reduce load after the school day is finished

Base load - reduced to circa 25 kWh overnight when the school is unoccupied

Further study

Energy Managers' Guide to Electricity Procurement from the Energy Manager's Association

A2.2 Roles and responsibilities

The energy manager, whether they have a technical background or not, has an increasingly important role in the organisation's bottom line - both from the financial perspective and from the environmental perspective.

Considerations for the energy management role

It is important that the energy manager does not feel isolated within an organisation. Energy management is everybody's problem; the energy manager is simply the key stakeholder with significant overall responsibility. However, they are ultimately only steering the ship.

Within large organisations, such as hospitals, there is often a fire officer. The role of the fire officer is to ensure within the estate the compliance of statutory requirements in respect to fire safety engineering and to provide advice to new projects. An organisation may decide that the energy manager's role may evolve to something similar.

Considerations for energy management responsibilities

The perception may be that the energy manager simply checks for lower energy supply costs. However, it should be understood that the act of balancing financial costs will involve reducing energy consumption, getting better energy return on energy invested in the system and hence reducing the implications of carbon based tax costs.

The energy manager will need to adopt a series of strategies, some of which will be related, some of which will be unique. However, the strategies will be seeking to reduce overall energy consumption whilst maintaining the same level of comfort and industrial output.

There are no single step solutions to energy reduction — multiple coordinated activities will be required.

It is important for an organisation to understand the roles and responsibilities of everybody on the site with respect to energy management. Every person interacting with an organisation and visiting its sites should have their own energy management responsibilities clearly defined.

	Self-assessment question: Are energy management roles and responsibilities defined?	Yes, or No?
a	There is no clear structure on energy management and the bill payer bears sole responsibility for energy use.	
b	There is some understanding across the organisation of individual roles. It is ad hoc and driven by individuals and does not necessarily match the energy management team's expectations.	
c	Energy management roles and responsibilities are clearly defined and properly delegated. They align satisfactorily with an agreed Plan of Work, the energy policy, and energy strategy requirements.	

The following two plan of work diagrams are based on the two halves of the energy management cycle outlined in Section 2.5:

(a) proactive energy management — purposes and principles; and
(b) reactive energy management — parameters and progression.

There are roles and responsibilities for all stakeholders at varying degrees of importance throughout the life cycle of the installation.

▼ **Figure A2.5** Plan of work diagram — purpose and principles

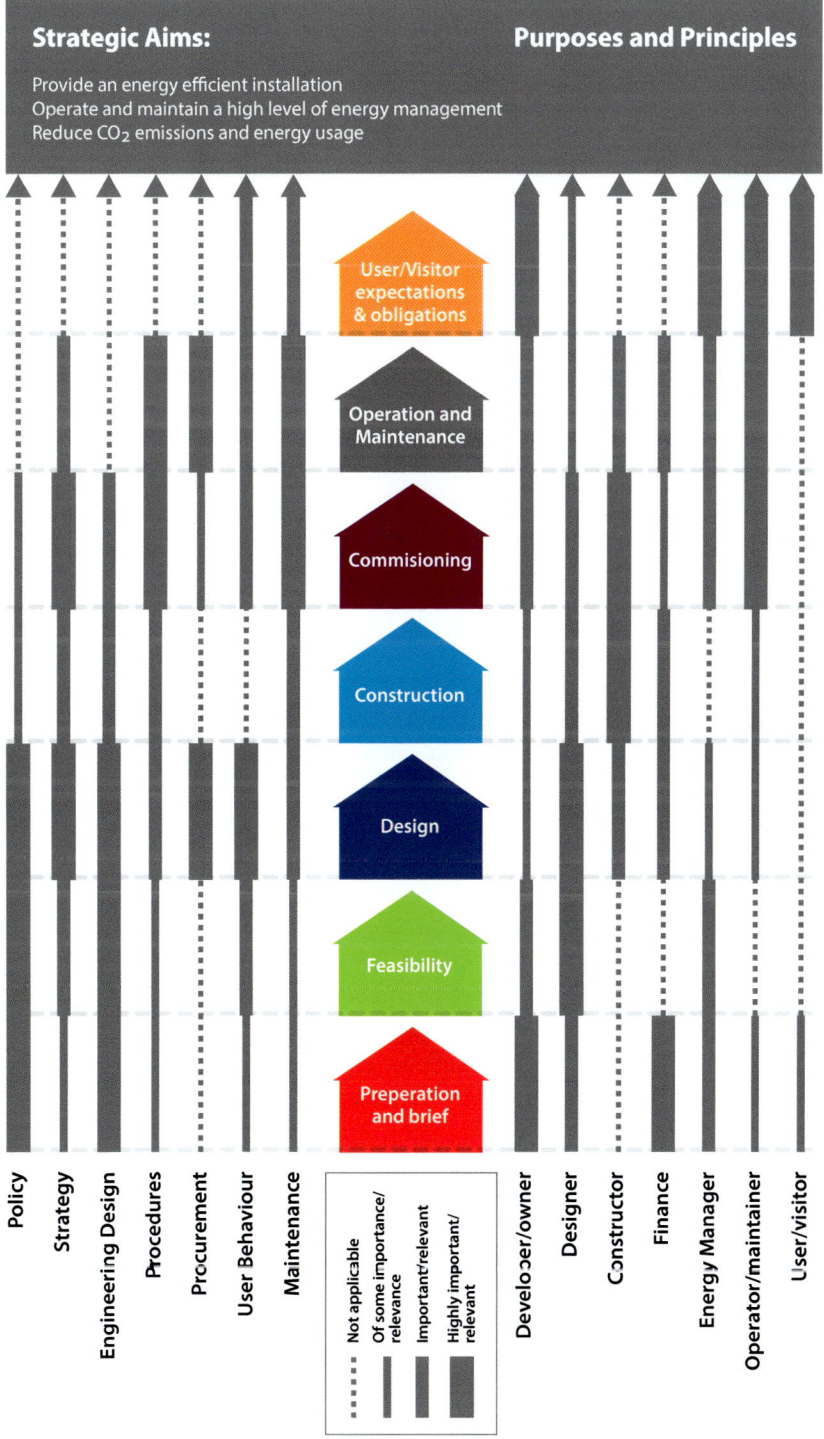

(Based on Plan of Work diagram from CIBSE/ADE CP1: *Heat networks — Code of Practice for the UK*)

Particular duties and responsibilities through the energy management cycle are described elsewhere in this Guide. Each of the seven main stages of the plan of work provides key roles for the various groups of stakeholders.

During the proactive stages of purpose and principles there will be overlaps, in their particular roles, of each of the stakeholders to fulfil the responsibilities from policy to maintenance. The three principal strategic aims at this point should be to:

(a) provide an energy efficient installation;
(b) operate and maintain a high level of energy management; and
(c) reduce energy usage and CO_2 emissions.

The early stages should create an operational environment that promotes and encourages energy management by all stakeholders. There will be greater emphasis on policy, strategy and design with involvement of the organisation's leadership team, designers and energy managers. Later responsibilities will include procurement and procedures and increased involvement of other stakeholders, such as contractors, and the operation and maintenance teams.

The energy manager has a role to fulfil in all stages of the plan of work with the possible exception of the construction phase. Whilst all the various responsibilities of the proactive half of the energy management cycle may not under the direct control of the energy manager, there will be liaison work necessary to ensure that the organisation's energy policy is being adhered.

Anticipating user behaviour will influence the later stages of the plan of work (for example at commissioning and operations/maintenance). Actual user experience, once the building is finally operational, will influence feedback and improvement plans in the reactive stages of the energy management cycle.

In the reactive stages there are still roles for all stakeholders. There will be greater levels of involvement for the energy manager, in particular, as performance is measured and improvements are set in motion. The three principal strategic aims should develop to be to:

(a) monitor and target an energy efficient installation;
(b) continue to operate and maintain a high level of energy management; and
(c) implement further reductions in energy usage and CO_2 emissions.

This will involve checking on use, comparing with benchmarks and setting new targets. Improvement projects may be instigated to provide step changes in the use of energy. Revisions to benchmarks may be necessary if the organisation's estate changes with either the inclusion of new installations, or the disposal of other assets.

The involvement of users and visitors, in shaping improvement projects and monitoring performance indicators, can assist in giving these elements a reality check. It also ensures that any energy saving measures, designed to improve performance, can actually work.

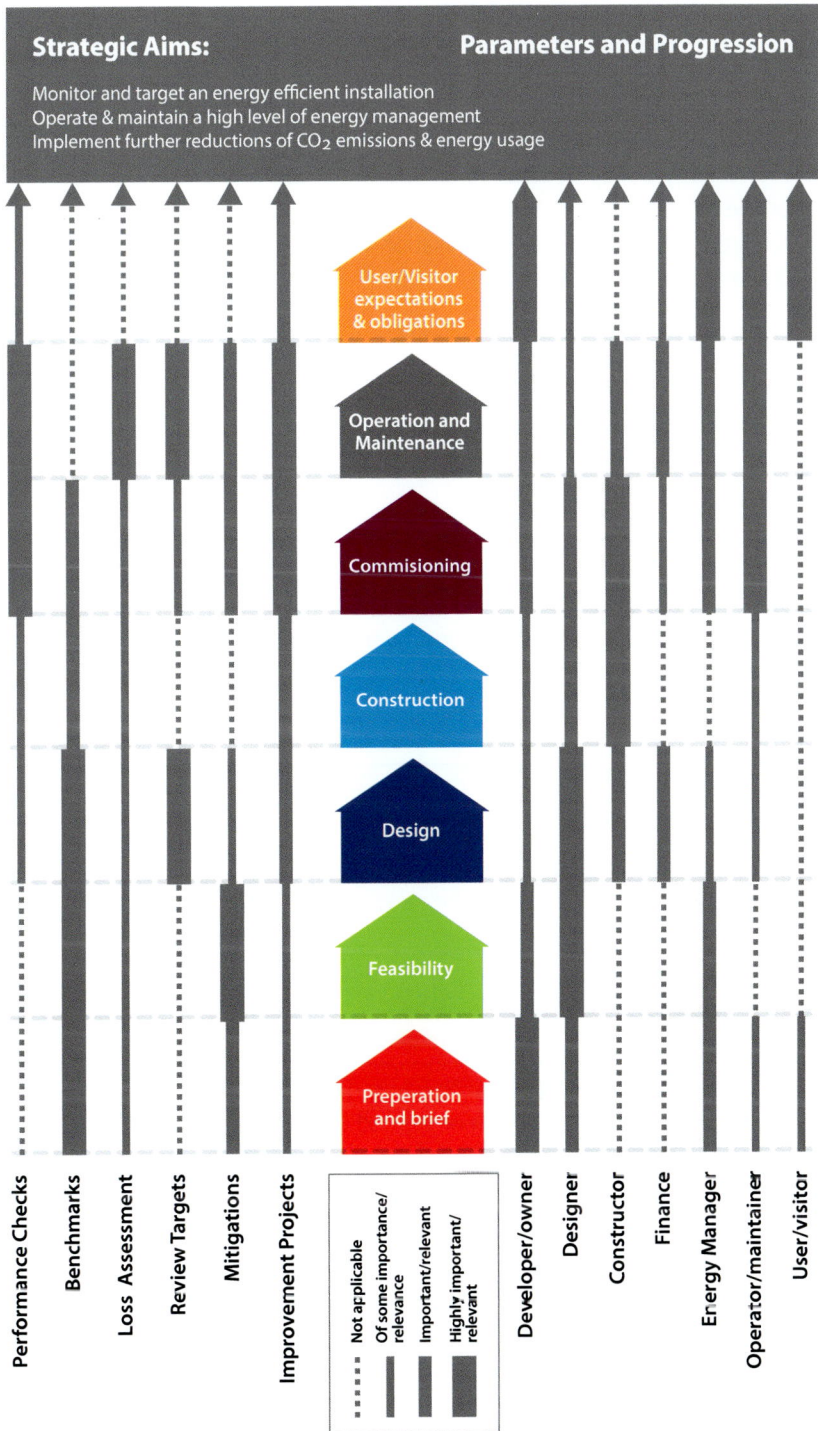

(Based on Plan of Work diagram from CIBSE/ADE CP1: *Heat networks — Code of Practice for the UK*)

There will be a number of key factors for the energy manager to understand, as well as the implications of those factors on the running of the organisation. Increasing energy costs can ultimately undermine business profitability, especially if business output does not increase.

Within an organisation it is important to establish:

(a) who is responsible for energy management policy and strategy;

(b) who is responsible for energy management and what technical resources are in place;

(c) who is responsible for monitoring energy use at all levels within an organisation;

(d) a communication system for energy policy updates to all stakeholders;

(e) a feedback communication system and reporting mechanisms for all stakeholders;

(f) what competences are required throughout the organisation and what skill gaps need to be filled;

(g) what training is required, at what level is it to be delivered and how it will implemented; and

(h) what additional measures are needed to facilitate cooperation and coordination throughout the organisation on energy related issues.

A2.3 User behaviour

A low cost but effective measure for reducing demand is to influence user behaviour. The introduction of active measures can encourage and inform user behaviour patterns so that activities within the built environment are energy efficient by default and not somebody else's problem.

Self-assessment question: Is user behaviour and associated patterns of use clearly understood?	Yes, or No?
a	There is no clear understanding or influence over user behaviour to avoid wasted energy consumption.	
b	There is some understanding across the organisation of user behaviour. It is driven by individuals and sometimes matches the energy management team's expectations.	
c	User behaviour is clearly understood and programmes to communicate and improve are in place. They align satisfactorily with an agreed plan of work, the energy policy, and energy strategy requirements.	

Getting all occupants, whether regular users or visitors, to actively participate in an energy saving ethos when on site can save significant amounts of energy. The Energy Managers Association launched an energy efficiency ratio in 2015, which described 40 % behaviour, 20 % controls and 40 % product as the primary focuses of attention.

It should be noted that the energy performance gap of new and refurbished properties is the subject of much debate. This is the lower measured actual performance of a building compared to the previous design performance predictions and targets.

When commissioning a new installation, it is important that users are adequately briefed on how the new controls are intended to be used to operate the building in an energy efficient manner. It is worth noting that installations with automated controls should also have manual overrides. Therefore, it is also important to explain to users the implications for energy consumption if these overrides are used.

The Carbon Trust document CTC 827 highlights seven stages to behavioural change based on knowledge, desire, skills, optimism, facilitation, stimulation and reinforcement. This should be cyclical to ensure continuous improvements.

The same document discusses four principal interventions in social marketing theory to influence behaviour; these could be developed and described as follows (Figure A2.7):

▼ **Figure A2.7** Influencing user behaviour

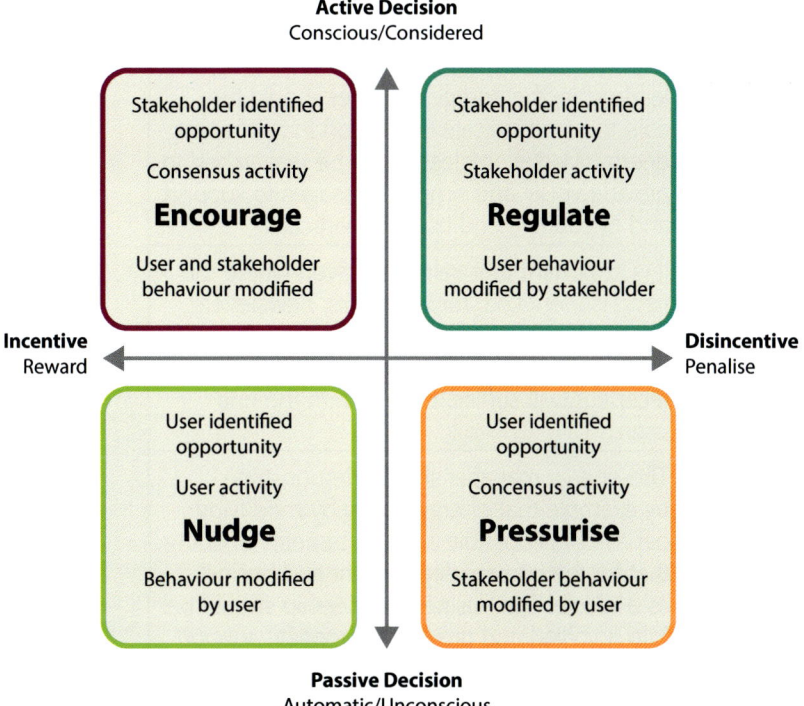

The energy manager may develop their own seven steps to influence user behaviour on energy management and provide feedback. This could be based on the following seven energy management related headings (Figure A2.8):

▼ **Figure A2.8** Step changes of user behaviour

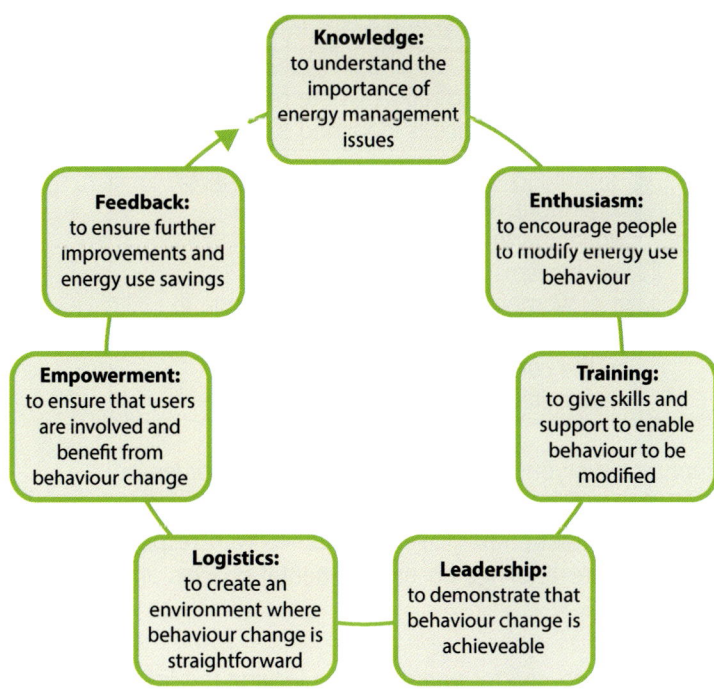

Aspect	Issue	Commentary
Knowledge	To understand the importance of energy management issues.	The energy manager needs to ensure that both end users and the business leadership fully understand the issues surrounding energy management and how it affects the business operations and viability.
Enthusiasm	To encourage people to modify energy use behaviour.	With the full support of the business leadership, the energy manager should be able to communicate the importance of saving energy. Marketing ideas could be researched to provide visual and simple messages to support and encourage end users in particular.
Training	To give skills and support to enable behaviour to be modified.	It is important that training is given where the need is identified. Tool box talks, newsletters, specific briefing can all be given. Do not assume that a simple notice on the board in reception will suffice. Engage with the end users.
Leadership	To demonstrate that behaviour change is achievable.	The energy manager should demonstrate by example that energy saving by modifying behaviour is possible and the benefits of doing that for the person, department and business as a whole. The business leadership should be fully involved and provide appropriate support.
Logistics	To create an environment where behaviour change is straightforward.	The energy manager should devise systems and procedures where changing the energy use behaviour is easy to achieve and becomes normal practice or second nature.
Empowerment	To ensure that users are involved and benefit from behaviour change.	The energy manager should encourage the end users to accept their own responsibilities for wasted energy use. End users should also be provided with: (a) mechanisms to address particular issues that have been identified by them; and (b) channels to report where reactive maintenance is identified as the solution.
Feedback	To ensure further improvements and energy use savings.	The energy manager should be looking for continuous monitoring of energy use. Working with users there should be mechanisms to identify areas for improvements and processes to change to reduce energy use where this does not impact on the commercial activities of the organisation.

Within the health and safety environment everyone is responsible, to a greater or lesser degree, for the health and safety of themselves and others depending on our roles. Sites often provide induction briefings and insist on particular procedures with respect to health and safety. Often linked to that are briefings on environmental policies and activities.

A similar ethos to safety norms should be adopted for energy use – leaving the lights on when an area is not in use is everyone's problem.

Measures, therefore, could include:

(a) site inductions to include key messages in a proscribed hierarchy of needs according to company policies – this should include health and safety, environmental considerations (waste disposal and water management), and energy conservation;

(b) clear posting of the organisation's key energy policies, alongside notices on health and safety and the environment;

(c) active regular feedback on periodic energy consumption, with clear comparisons between targets and actual performance in a graphical easily understood format;

(d) training of specific duty holders with respect to local energy management at departmental level with clear terms of reference for the roles; and

(e) feedback routes to ensure two-way communication to the correct channels on energy management related issues.

A2.4 Maintenance

Engineering designs should consider how the installation will be maintained for ease of access and also to ensure that sustainable operation continues in line with the original design concepts. Standards such as BS EN ISO 50001 can provide guidance and help inform this part of the process. Other tools for auditing, energy reporting and fulfilling government requirements can also be used.

	Self-assessment question: Are communication channels in place between energy managers and operational maintenance teams?	Yes, or No?
a	There is no link up between energy managers and maintenance teams. Priorities for energy efficiency in design and operation are not coordinated.	
b	There is some understanding across the organisation of the link between maintenance and energy management. The channels are driven by individuals rather than strategy. It sometimes matches the energy management team's expectations.	
c	The link between maintenance and energy management is clearly understood and programmes to communicate and improve are in place. They align satisfactorily with an agreed Plan of Work, the energy policy, and energy strategy requirements.	

The energy manager should liaise with the maintenance department management to ensure performance reliability and efficiency is maintained throughout the life cycle of the installation. Any sustainable design that is not easily maintained will quickly become unsustainable – any energy saving measures will be negated and the equipment itself decommissioned.

Robust maintenance regimes have an important role in an energy conscious and more sustainable world. There can be no doubt that correctly maintained engineering systems will operate at their maximum energy efficiency for much longer.

Energy managers should undertake regular energy audits. These audits should be linked to a maintenance audit with specific reporting of sustainable operations of green technology.

As technology advances maintainers and asset managers will increasingly utilise automated means of monitoring performance and computer systems to plan, manage and report on maintenance activities. Regular reviews of these maintenance systems should form part of an energy manager's regular energy management audit process.

The following case study first appeared in BRE Information Paper IP7/13 and is reproduced by kind permission of the original Author Dr Andy Lewry of BRE.

The BRE information paper in general, and the case study in particular, highlights the importance in understanding the principles of energy management and of using timely and accurate data to move the agenda forward.

Like many other museums and galleries, Merseyside Maritime Museum had not taken up the provision for 30-minute data readings for its electricity, gas or water consumption from the utility companies. This meant that it was not possible to pinpoint what was causing noticeable variations in energy use.

An analysis of half-hourly data highlighted that 400 kW of electricity was typically used during the day and that some 280 kW was still being used at night. It identified that the chillers, air-handling units (AHUs) and heating systems were all in operation 24 hours a day, seven days a week, in order to maintain the appropriate temperature level and a relative humidity of 50 % to preserve the museum's artefacts.

It was recommended that the museum should change the operation of the three AHUs serving the theatre, entrance foyer and shop. Time switches were installed, resulting in an automatic switch-off whenever the building was unoccupied.

To investigate how the chillers should be operating, additional sub-metering was installed on all three chiller units. The system performance was changed by relaxing the chiller controls and their set points – setting the bands to 40-60 % instead of 50 % (where appropriate). The next step was to install temperature and humidity sensors to measure the external conditions and observe how effective the systems were.

This has resulted in substantial electricity, gas and water savings for National Museums Liverpool. The gas consumption has been reduced by 25 % and electricity consumption by 7 %; in seven months this equated to savings of £22,633.

This case study was reproduced by kind permission of TEAM (Energy Auditing Agency Ltd).

Managing performance, benchmarks and losses

Having set the purpose and principles in place, the next steps are to take stock and evaluate the results of the proactive half of the energy management cycle. These reactive tasks will involve a combination of responses to check the performance of the installation and the energy consumption overall, to validate existing benchmarks and update where required and to evaluate disparities and losses in performance.

A3.1 Performance checks

It is important that the organisation implements an effective auditing strategy to monitor:

(a) the energy performance of the installation overall;
(b) periodic energy costs;
(c) trends in energy consumption;
(d) progress on loss mitigation; and
(e) the activities of staff and end users to assist with energy management at a local level.

	Self-assessment question: Is there a process in place for systemic evaluation and auditing of the energy consumption?	Yes, or No?
a	There is no energy auditing process in place beyond that of checking and paying the energy bills.	
b	There is some understanding of the need to periodically check site wide energy consumption. It is driven by individuals rather than strategy. It sometimes matches the energy management team's expectations.	
c	There is a clear understanding of the need to regularly check energy performance at a local level and also site wide. Processes are in place. They align satisfactorily with an agreed Plan of Work, the energy policy, and energy strategy requirements.	

This auditing strategy should be managed carefully, carried out at regular intervals and periodically reviewed to ensure it is fit for purpose and current needs.

The general requirements for energy audits are described in BS EN 16247-1. Further information on particular systems are covered in:

(a) Part 2 (buildings);
(b) Part 3 (processes); and
(c) Part 4 (transportation).

The competence of energy auditors is discussed in part 5 of BS EN 16247-1.

Increasing levels of government led legislation will compel organisations to declare headline energy performance criteria on an annual basis. Organisations who proactively manage their energy consumption will fulfil the needs of those basic reports and also demonstrate that they are properly auditing their energy management systems.

It should be recognised that support at board level within an organisation is critical to the success of auditing energy management systems and to the success of using the results to initiate further improvements.

There is also a need to understand what is actually important to the business. Inhibiting energy consumption in one part of the infrastructure may provide energy savings locally but could also result in loss of production in another part. It could also result in increased consumption on another part of the installation because of overcompensation. Passing the problem for one part of a production system to another, as an example, without looking at the whole system holistically will not achieve very much.

The energy manager should understand and guide the organisation on how to monitor the various parts of the energy infrastructure from the metered intake through the distribution system to the point of use. Strategically placed sub-meters will ensure that meaningful data can be collected. These meters can be readily used to monitor and subsequently improve energy consumption.

Meters and sub-meters are important tools in the monitoring of energy consumption and will assist in informing energy management techniques. However recognised guidance should be followed to ensure the correct meters are specified and that they are placed in the optimum location. Careful deployment of metering will enable the energy manager to understand what is being used and where.

Documents from CIBSE and IET will assist with formulating a strategy to implement metering and sub-metering and also understanding the requirements for non-fiscal energy metering. These include:

(a) CIBSE TM39 *Building Energy Metering.*
(b) IET *Guide to Metering Systems, Specification, Installation and Use.*

The energy manager should also ensure they, their team, and their colleagues in maintenance, can read, understand and correctly interpret what the meters are reporting.

Careful and informed analysis of meter outputs is required. This is especially true of electrical meters that report on not just on power demands (kW) and energy consumption (kWh), but also on power factors, harmonics and other power quality aspects that directly affect the operational efficiency of the system.

Regular performance checks from strategically placed energy meters will enable timely and accurate analysis so that:

(a) normal energy patterns of use can be monitored and reviewed at appropriate intervals;
(b) abnormal energy events can be highlighted for further assessment as soon as is practicable;
(c) power quality assessments on electrical infrastructure can be monitored so that appropriate corrective measures can take place; and
(d) fiscal reconciliation can be made on the correct type of meters so that billing of tenants with appropriate meters can be made.

IEC 60364-8-1 describes a process that electrical installation designers can use to differentiate energy consuming loads according to their usage or load characteristics, the geographical zones in which they operate within an installation and the meshes where appliances may share control philosophies and technology, sometimes across multiple zones.

An IET publication, *Designers Guide to Energy Efficient Electrical Installations*, references IEC 60364-8-1 and shows how designers can use this process. It demonstrates also the responsibilities of clients and end users in developing energy efficient design solutions and the process for setting out benchmarks for performance during the design stage.

To illustrate the value of understanding energy usage within a wider installation, the percentage values in the following graphic have been provided by Salix Finance Ltd. This publicly funded organisation facilitates energy saving opportunities. Their work has demonstrated the following typical loads, or energy usage for most schools. A similar approach can also be applied when profiling carbon and financial usage of energy within any installation or estate.

▼ **Figure A3.1** Typical energy consumption figures for schools (Based on information from SALIX)

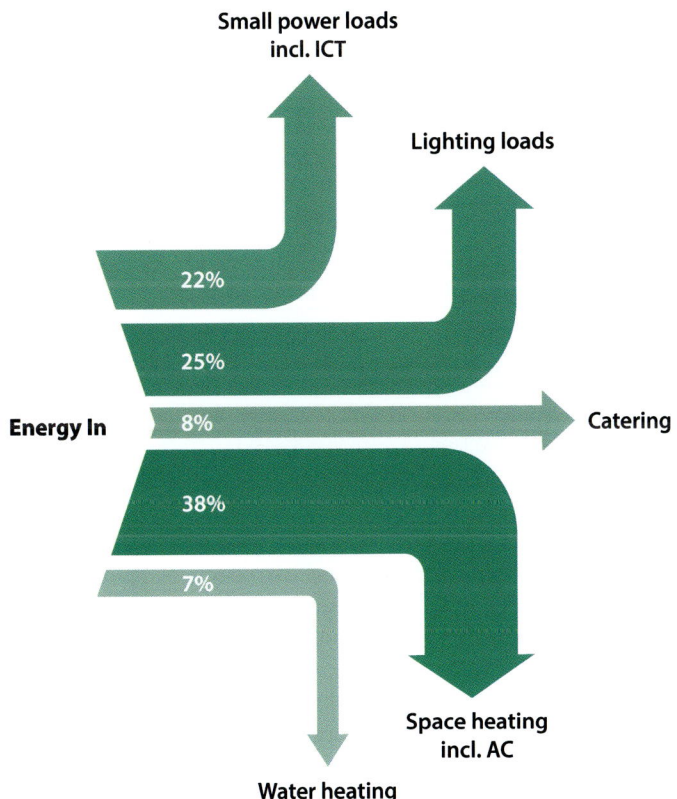

Understanding energy consumption at this level of detail will provide guidance on where to focus attention with energy saving projects.

Greater lighting controls (daylighting and/or occupancy controls) and replacement LED lighting could yield significant savings on the typical 25 % overall energy consumption in a relatively cost effective manner.

Similarly, space heating strategies again focussed on occupancy controls and also heat recovery could yield satisfactory results but may be more complex to deliver.

Catering and water heating will need some specialist attention and may only produce marginal energy savings.

Information and communications technology (ICT) is an increasingly important part of business life. Careful management of the systems to allow a business to function and flourish is important. Close management of the development of these systems is required to ensure the energy used is carefully managed.

Often too many devices are crammed into hubs that then need cooling by external means. Poor cable management in hubs also restricts air flow and reduces cooling capacity. Switches that run at high temperatures will fail prematurely. How important is the ICT system to the business?

A principal task of the energy manager will be to double check meters and verify energy bills.

Check	Commentary
Check energy types	(a) Are all energy types (gas/electricity/heat networks) clearly identified? (b) Are all intake energy meters clearly identified and locations mapped across the site? (c) Do all of the fiscal meter identity codes correlate with the relevant energy bills and statements?
Check fiscal energy bills	(a) Are all elements of the bills clearly understood? (b) Does the capacity charge reflect current and future needs? (c) Are the peak consumption values noted? (d) How does the recent period compare to the preceding period and also the same period in the previous year?
Check fiscal meters	(a) Where possible check and verify energy consumption readings of the main intake meters and record values.
Check sub-meters	(a) Check and verify energy consumption readings of all sub-meters and record values. (b) Summate sub-meter readings and compare to main intake reading. (c) For electrical sub-meters, check power quality values and record. Readings should be analysed to ensure energy efficiency parameters are not exceeded. (d) Check the location of all sub-meters. Are more required to give more granular levels of information? (e) Consider use of sub-meters for separate zones, such as plantrooms, sub-letting areas, separation of offices and process areas in factories and warehouses. (f) Consider use of separate mesh based sub-meters for small power and lighting especially in common areas of large complexes.

Regular updates of checklists and asset management plans will ensure complete understanding of where is energy used within the business from the point of supply through the distribution infrastructure and on through the process.

The energy manager should consider the following examples, and working with other site stakeholders, develop other areas for consideration. Where necessary technical staff or contractors should be delegated to carry out site surveys to ensure the engineering systems provide the optimum energy performance levels.

What	Commentary
Supply intakes	Are they of sufficient capacity for existing requirements? Are they placed in the optimum location to reduce infrastructure losses (barycentre design philosophy)? Has additional allowance or planning been made for future needs?
Infrastructure	Is the services infrastructure correctly sized? Has it been installed correctly with minimum run lengths? If insulation is required (pipework, ductwork etc.), has this been checked for gaps or damage recently?
Site services	Does the installation still meet current needs? Have future developments been considered with the impact on site services? Typical considerations will be mechanical processes such as steam or compressed air. Other considerations will be mechanical environmental services such ventilation, space and water heating, space air conditioning and cooling.
Site boundaries	What site areas are owned and operated by the business with direct control over energy consumption? What site areas are leased and therefore the business has less control over in terms of energy management? Does the business have influence over these leased areas for a coherent energy management operation for the whole site?
Public areas	Does the area have sufficient illumination to allow safe entry and egress? Is sufficient use made of daylighting to reduce use of artificial lighting at appropriate times? Is sufficient use made of heating and cooling temperature controls in public areas? Is the controls strategy for the ventilation correct for the use of the space?
Offices	Does the area have sufficient illumination to allow safe entry and egress? Is sufficient use made of daylighting to reduce use of artificial lighting? Is sufficient use made of occupancy controls to set back lighting (dim), or switch off, when not in use? Is sufficient use made of heating and cooling temperature controls in public areas? Is the controls strategy for the ventilation correct for the use of the space?
Site utilisation	Can under-utilised or unoccupied areas be controlled to minimum flow rates (space heating) or isolated completely (mothballed installations)? What are the environmental implications of not heating an empty space especially in cooler months?
Plantrooms	Have back of house areas been maintained sufficiently to ensure continuing energy efficient operations of plant equipment, pipework and ductwork? Are regular inspections made and reports acted on as required?

What	Commentary
Production machinery and equipment	Is there sufficient understanding of how factory based production machinery and equipment uses energy Have processes been analysed for energy savings? An example might be a large biscuit making line which has a number of heated sections. Do they all need to be on all the time if at differing process stages? If individual sections are isolated, what is the reheat time and will it affect production? Is there potential for heat recovery? Is there sufficient understanding of operating duties of equipment, plant or processes? Is equipment left on but not used for long periods?
Resilience	What is the impact of failure? In a disaster recovery scenario can energy consumption be controlled and managed to minimise higher tariffs? Or is the cost to the business more if production halts? If there is a reliance on the use of renewable site supplies or low carbon alternatives (CHP), then is there a back-up plan when these are off line for routine maintenance or not operational because of engineering failures? What are the requirements for resilience of relevant energy use across the organisation, such as the need for standby or backup supplies?

A3.2 Benchmarks

The energy manager should be aware of the expected performance benchmarks for the building:

(a) for new buildings this will be the design data and the initial commissioning measures on practical completion;

(b) for older buildings the initial design data should be updated with seasonal data in subsequent years for direct comparison; and

(c) recent refurbishment projects and developments will also inform and update the site wide energy performance parameters.

The operations and maintenance data for the site will be an important reference document. The key stakeholder for the energy manager to consult with will typically be the estates maintenance team. For smaller sites consultation with relevant contractors and consultants may be necessary.

	Self-assessment question: Does the energy management team understand what benchmarks are in place and are they reviewed to accommodate change of use to buildings?	Yes, or No?
a	There are no existing benchmarks in place to meaningfully evaluate the current performance of the building.	
b	There are some benchmarks in place but they have not been reviewed or updated. Building performance sometimes matches the energy management team's expectations.	
c	There is clear understanding of the commissioned building performance benchmarks. These are regularly checked and updated where appropriate. Processes are in place. They align satisfactorily with an agreed Plan of Work, the energy policy, and energy strategy requirements.	

It is important that the energy manager sets up regular reviews of the expected performance benchmarks. The reviews need to take into consideration some of the following:

(a) the original design data;

(b) any submissions to award bodies such as BREEAM, LEED etc.;

(c) the current use of the facilities;

(d) any changes to the overall footprint; and

(e) any changes of use within an area of a particular building.

Caution:

The correct level of detail is vitally important throughout the installation to make informed judgements on patterns of energy use. Just analysing the main intakes is not sufficient. Strategically placed sub-meters enable benchmarks to accurately compare to real-time energy data.

An example of this may be the sudden increase in uncontrolled energy consumption in one area of the business that might be offset by the coincidental decommissioning of another area. When the decommissioned area comes back on line after refurbishment the focus on any problem solving with regards to energy performance could be incorrectly focussed on that area and not on the actual original problem.

With the level of granular detail afforded by sub-meters, more accurate analysis and judgements can be made.

The energy manager should work with other business stakeholders to identify other changes to energy use patterns such as:

(a) new process equipment within factories;

(b) larger replacement plant than previously installed;

(c) changes in work patterns — such as moving to double or triple operating shift pattern; and

(d) changes in occupancy levels — such as new staff on a previously unoccupied floor.

The energy manager should seek out industry guidance related to their particular type of installation as a comparison to see how their own estate is performing. Documents, such as CIBSE Technical Memorandum TM46 on Energy Benchmarks, provide comprehensive guidance notes and tables of data for various types of installation and levels of use.

Making sense of the comparative values of design data benchmarks and actual consumption data is a complex task. For energy managers not familiar with the formula and processes involved, it is advisable to seek advice or engage the services of an external consultant with the appropriate skills and knowledge.

Submission of energy consumption results to national schemes, such as ESOS, is already a necessity for larger enterprises. Such reporting mechanisms are likely to involve smaller enterprises at some point in the future. Accurate reporting of energy benchmarks based on recent data sets will become part of the energy mangers remit.

Among various criteria that the energy manager will need to familiarise themselves with is:

(a) the comparative on-site use of electrical energy and of fossil fuels;

(b) the importance of degree days analysis in understanding the efficiency of space heating use as seasonal external temperatures vary;

(c) understanding how to accurately calculate power consumption ratios in a common unit like kWh/m^2;

(d) how to use calculated kWh/m^2 values to convert to kg/CO_2 for reporting purposes and analysis;

(e) the necessity to separate specific energy loads for more accurate building calculations;

(f) potential energy consuming loads or processes to omit from calculations or to treat separately;

(g) how to address the needs of multi-use buildings with varying energy consumption; and

(h) the importance of sub-metering different types of energy consuming loads and timely compilation of readings to accurately report trends, assess efficiencies and highlight potential losses.

A method of calculating energy performance for evaluation with previous years and other similar buildings is asses the normalised performance indicator. This gives a value in kilowatt-hours per square metre (kWh/m^2).

A simplified flow chart is shown below in Figure A3.2. The method is described in detail in various academic text books.

Further study

Chapter 2, Energy Audits in Buildings

Energy Audits by Tarik Al-Shemmeri (ISBN 978-0-4706-5608-2).

▼ **Figure A3.2** Benchmarks flowchart

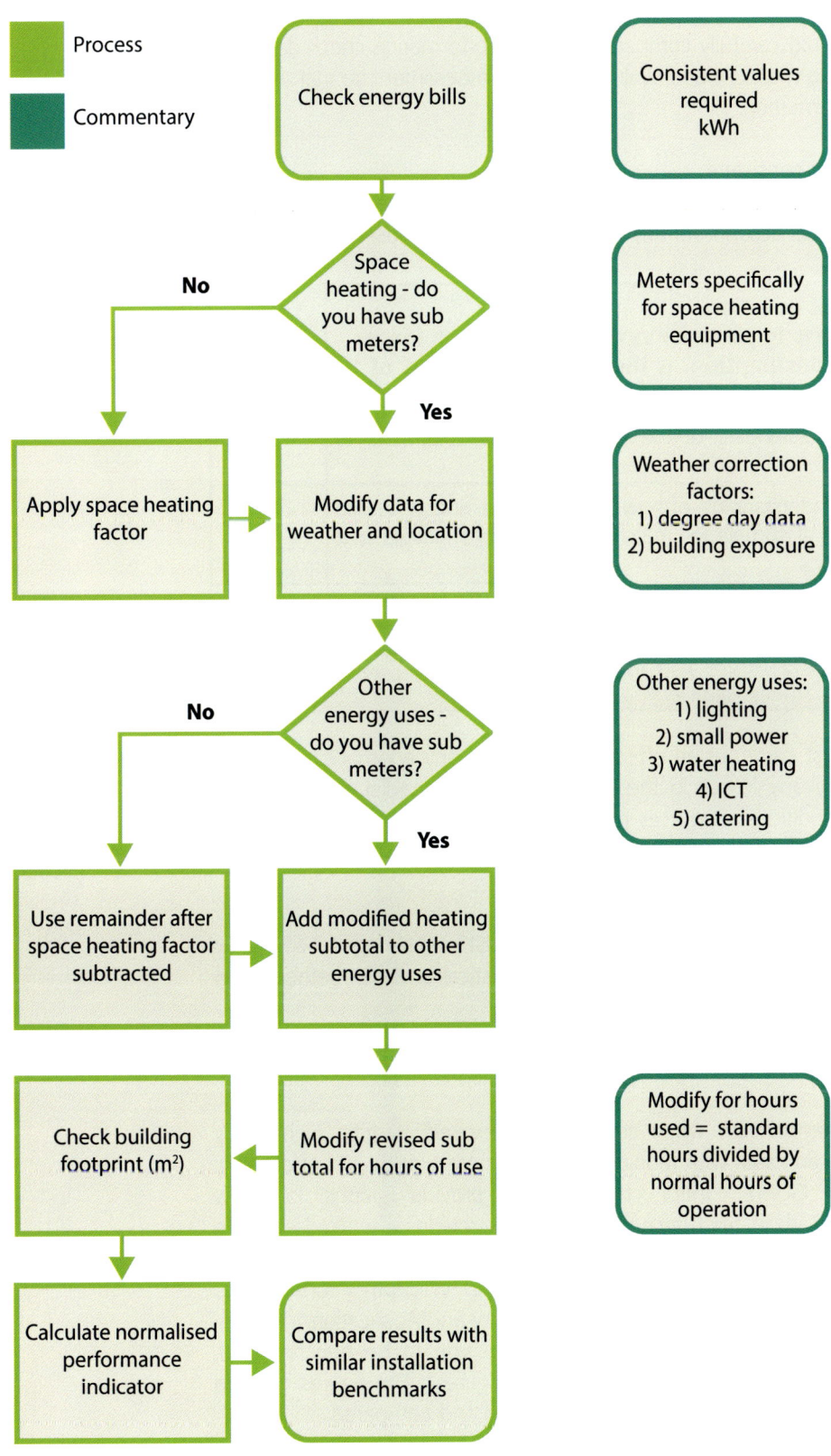

A3.3 Analysing losses

The energy manager should carefully compare both the performance check data, and the known benchmarks, to reconcile and evaluate any loss assessment so that they can be confident that the information:

(a) from both data sets is consistent;

(b) illustrates any new and evolving energy use in the building accurately; and

(c) highlights and takes into account for any adverse events or trends.

	Self-assessment question: Is there a process in place for systemic evaluation losses in the energy consumption? Is there a regular review of the installation?	Yes, or No?
a	There is no understanding of the losses nor available skills to assess them.	
b	There is some understanding of how to assess the energy losses in the installation. It sometimes matches the energy management team's expectations.	
c	There is clear understanding of how to regularly check the installation for energy performance losses at a local level. Processes are in place. They align satisfactorily with an agreed Plan of Work, the energy policy, and energy strategy requirements.	

The energy manager must carry out an effective analysis of any anomalies in energy performance and identify any losses in the system so that remedial action can be considered. The use of additional resources may be prudent here, including energy or engineering consultants, to analyse trends, or spikes in energy use, and hence isolate potential causes of losses.

For space heating the use of degree day calculations will assist in ensuring that a building is largely operating within normal parameters. Further statistical analysis methods may help to highlight particular problems:

(a) Regression analysis normalises past energy performance dependent on the relative weather data at the time.

(b) Exception reports demonstrate when consumption exceeds targets highlighting wasted energy. This can assist with monitoring usage more quickly and effectively.

(c) Cumulative sum (CUSUM) control chart reports provide graphs to show performance trends. Carbon Trust document CTG077 provides a worked example for the use of CUSUM and how the tool can be deployed to monitor, analyse and subsequently improve energy performance. Further in depth information is available in CIBSE document TM41 Degree Days: Theory and Application.

(d) Key Performance Indicators (KPI's) allow performance you to be benchmarked. This uses key metrics (e.g. floor area or production figures). The figures collated can be used for tables that can be periodically updated and compared.

For electrical loads, up to date checks on the load schedules should highlight if new equipment or buildings have been added. Regular checks, using appropriate placed and specified meters and sub-meters, will also allow power quality analysis such as poor power factors, reactive impedance loads and adverse harmonics to be spotted.

Load profiling of the main intake meters, and where necessary at principal sub-meters, at regular intervals throughout a 24-hour 7-day cycle will allow:

(a) normal energy use to be monitored;
(b) acute adverse events to be highlighted as soon as possible;
(c) adverse, but slow changing, trends to be investigated; and
(d) improvements to be assessed and targets set.

The three most likely main areas of potential energy losses will be:

(a) fabric:
   i   through the external façade of the building itself; and
   ii internally with loss of heat through poor insulation on the infrastructure such as pipework or ductwork.

(b) engineering:
   i   failure of plant controls or infrastructure leaks that require reactive maintenance to remediate; and
   ii poor performance of equipment or systems due to inadequate preventative maintenance.

(c) people:
   i   through poor understanding of energy procedures;
   ii   leaving appliances in use when unattended or when not actually required;
   iii manually overriding automated controls;
   iv   leaving external doors open during the heating season allowing heat to escape; and
   v   leaving external doors open during the cooling season making air conditioning work inefficiently.

Case Studies

The Carbon Trust publication, CTG077 Monitoring and Targeting, highlights the following case studies to illustrate why deployment of constant monitoring and targeting is advisable and checks necessary to catch losses in real time. Each of these cases could have been avoided by more care in maintenance and better information of installed systems:

(a) Losses that would have equalled £3,500 a year were incurred when a limit switch came loose at a waste-water treatment works, causing some machinery that should have been running intermittently to run continuously.
(b) Gas consumption at a council depot doubled when a maintenance contractor left the heating system running 24 hours a day.
(c) The front steps of an office block were found to be costing £5,000 a year because the control had failed on the de-icing heaters (which the owners did not even know they had).
(d) Frost-protection systems are a prime cause of waste. One energy manager cut the electricity consumption of his HQ building by 40 % when he found that all the electric frost protection pre-heaters were running on his air handling units.
(e) Hundreds of pounds a year were lost when someone left a bypass valve open on a steam trap in the basement of a paper mill.

The energy manager should be aware of the slightly different approaches, and hence outcomes, expected of:

(a) monitoring and targeting; and
(b) measurement and verification.

Monitoring and Targeting

The Carbon Trust publication CTG 077 defines monitoring and targeting as follows:

"The purpose of monitoring and targeting (M&T) is:

(a) to enable an understanding of your energy consumption data;
(b) identify underlying factors which impact upon consumption;
(c) and set appropriate targets that allow you to review performance.

This will subsequently enable you to identify avoidable waste or other opportunities to reduce consumption."

With appropriate understanding of the engineering systems in place and knowledge of the consumption levels, monitoring and targeting can be conducted in house and the information used internally within an organisation to improve performance.

Measurement and verification

Initially, measurement and verification (M&V) determines the method of measuring the savings, firstly in energy terms and then in financial terms, of any improvement projects. It shows that tangible results have been achieved for energy manager reports and also for the finance department.

Additionally, it demonstrates continuous improvement by the organisation. A supporting document to ISO 50001, ISO 50015, specifically provides information and guidance on the general principles for measurement and verification of energy performance.

M&V typically uses particular measurement strategies and the International Performance Measurement and Verification Protocol (IPMVP) outlines four options:

(a) Option A — retrofit isolation to enable measuring of key parameters.
(b) Option B — retrofit isolation to enable measuring of all parameters.
(c) Option C — measuring the entire building or whole facility.
(d) Option D — calibrated simulation.

Understanding the data is a complex task and involves accurately analysing the statistical information to correctly interpret:

(a) internal influences such as:
   i   changes in building use;
   ii   new additions to the estate;
   iii   user behaviour or; and
   iv   the processes that take place.

(b) external influences such as weather patterns.

Aspects such as legislative compliance, correctly assessing energy performance and financial returns mean that there is a need to verify the data to ensure accuracy. The measurement and verification plan needs to be in place first and developing that needs careful consideration.

Large estates and buildings may be in a constant state of flux as various improvements are carried out. The different programmes of activity may not be synchronised. Therefore, any measurement and verification process needs to cope with these constant changes to make the resultant data meaningful.

It is recommended that specialists are engaged to assist with measurement and verification, especially with large organisations that have complex installations. Specialists in this area are usually professionally accredited as Certified Measurement and Verification Professionals (CMVP).

Further study

(a) Carbon Trust CTG077 *Monitoring and Targeting*
(b) Carbon Trust CTG075 *Degree Days for Energy Management*
(c) CIBSE Technical Memorandum TM41:2006 *Degree-days: theory and application*
(d) Appendix C of this Guide provides detail on specific engineering services and potential losses.

The following case study first appeared in BRE Information Paper IP7/13 and is reproduced by kind permission of the original Author Dr Andy Lewry of BRE.

The BRE information paper in general, and the case study in particular, highlights the importance in understanding the parameters of energy management and in working with data to understand energy losses and identify priority areas for improvement projects.

In 2000, City College Plymouth adopted a target to reduce its carbon emissions by 34 % compared with 2001. The college has made excellent progress, as is shown in Figure A3.3.

Figure A3.3   City College Plymouth carbon emissions

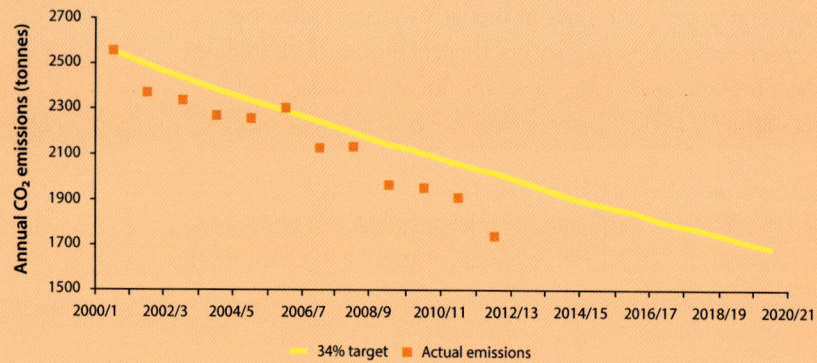

This shows that since the year 2000 emissions have been reduced from 2554 tonnes to just 1740 tonnes, a reduction of 31.8 %, and putting the college very close to the 2020 target. However, there has not been any relaxation in effort, and additional opportunities are actively sought through:

(a) detailed analysis of energy use data;
(b) a continual process of improving metering and information retrieval; and
(c) specific process and building investigations.

Originally, the main site used oil for heating, this being fired in three 2 MW shell and tube boilers operating a medium-temperature system. Although combustion efficiencies were relatively good, the size of the plant made it very inefficient at low load factors. Meters installed on the hot-water system proved that big savings could be made by improving the efficiency of domestic hot water production.

The installation of a 60 kW gas-fired water heater in place of the original 600 kW calorifier had a dramatic effect on fossil fuel use, cutting the cost in the summer months from £2000 to £200, allowing the main boilers to be shut down for about four months of the year. This measure had a payback of less than one year.

At an early stage it became apparent that plant control was poor. Things improved once it could be controlled locally on site rather than at the local authority offices. Among other advantages, this allowed effective scheduling of plant, including the six large air-handling units that previously ran continuously with an aggregate electricity demand of 60 kW, a figure identified through the use of an energy analyser. This single measure reduced electricity use by about 300,000 kWh per annum.

Further savings were achieved through the installation of 'inverter' variable speed drives; this enhancement not only further reduced electricity use but also reduced thermal demands as ventilation rates were reduced from the previous 12 air-changes per hour to less than three.

More recently, the original 1970 boiler installation has been replaced by four 500 kW condensing boilers, with compensation of the boiler temperatures according to the weather. Much of the site has also benefitted from the installation of cavity wall insulation. The main tower block needed replacement windows, and, during this process the building was also clad with 100 mm of insulation and fitted with a *brise-soleil* shading system.

The main site's two 1000 kVA transformers have been 'tapped down' from circa 250 V to 235 V, and consideration is being given to replacing the transformers with new low-loss units more appropriate for the site's reduced electricity demand of about 750 kVA.

In the School of Catering, electric induction hobs have replaced many of the traditional gas burners.

City College Plymouth expects to continue to be a leader in energy efficiency through scientific investigation and good engineering practice until 2020 and beyond.

This case study was reproduced by kind permission of City College, Plymouth.

Managing reviews, mitigations and improvements

Building on the reactive philosophy, the progression of energy management should provide a comprehensive analysis of recent energy usage and set the next cycle of energy management activities in motion.

A4.1 Review of targets

The energy manager should periodically review targets to ensure that:

(a) existing targets match the current capacity of the installation;
(b) targets are revised down for installations that are reducing in size;
(c) targets are revised upwards for installations that are expanding in size;
(d) any new developments are catered for within any revised targets; and
(e) timescales for implementation of the new developments are understood so targets can be adjusted accordingly.

	Self-assessment question: Is there a process in place for systemic evaluation of energy consumption targets? Does the process include adjustments where required?	Yes, or No?
a	There are no energy targets in place.	
b	There is some understanding of energy targets for the installation and how to monitor them. It sometimes matches the energy management team's expectations.	
c	There is clear understanding of how to evaluate energy targets. Monitoring technology and processes are in place. They align satisfactorily with an agreed Plan of Work, the energy policy, and energy strategy requirements.	

Energy management systems should acknowledge that there will be a series of iterative steps each achieving small gains on the road to overall energy reduction. These steps should be continually monitored to seek further reductions and all improved where it is reasonably practical.

Whilst the setting of targets is advisable to provide focus in energy management systems, these targets need to be realistic to enable business growth, and adaptability to change. Any targets need to be reviewed against the overall energy strategy of the business.

The energy manager should be aware that setting targets too leniently may cause unique abnormal events or adverse continuous activity to be missed until they become a larger problem. Likewise targets that are too aggressive will affect the morale of the end users to the extent that energy conservation does not feature on their priorities.

Energy management is everybody's problem, but a successful energy management system needs engaged users. Do not disenfranchise users with unrealistic targets.

Within the section entitled "Estimating expected consumption", CTG077 describes a number of methods to collate previous performance. These methods take the raw data from the energy monitoring function and take into account extenuating circumstances to explain variability and driving factors behind energy consumption such as heating degree days, cooling degree days and also output (useful for industrial installations). These include:

Activity	Commentary
Precedent forecasting	Commonly used on monthly monitoring schemes, this methodology is based largely around noting the respective period on the previous year and using that as the target. This simplistic method should only be used as a guide, especially for space heating. It is prone to errors because of changing temperatures caused by unpredictable weather patterns. Abnormal energy consumption, especially one off events, can also make this forecasting method inaccurate. For relatively constant loads in relatively stable environments this method can provide reasonable target estimates if the results are treated with caution.
Activity based targeting	Various methods can be used to predict energy. There may be a variety of driving factors, usually based on external factors, that cannot be controlled, but will still influence energy consumption on site. Examples of driving factors will include: (a) outside temperature affecting space heating; (b) humidity affecting air quality, ventilation and air conditioning; and (c) exterior light levels affecting use of lighting.
Consumption driven by the weather	External temperature and weather patterns will affect the reliance of occupants on either increased space heating requirements or conversely on cooling. Predictions and targets can be predicted using degree day charts but reliable outcomes are dependent on engineering controls.
Consumption driven by production throughput	Planned increases in production should provide information that can be correlated against previous energy performance to provide reasonable indicators of energy consumption. Similar charts, based this time on output measured in tonnes, can be used to predict likely energy use.
Alternatives to straight-line targets	Potentially used for factory production lines, this method relies on previous energy readings for known quantities of products. Once found, the pro rata energy value in kWh can then be factored up or down as the planned quantities of produced units increase or decrease.

The energy manager should be familiar and comfortable with compiling the correct information into graphs and using this information for further analysis. These tools will highlight typical energy performance levels in a comparable way.

Updating the graphs with real time data will enable the energy manager to highlight abnormal events at the earliest opportunity.

The graphs can also be used to predict probable energy performance targets as the size of the installation or estate changes.

As an aid to the energy manager, information from the maintenance and operations teams on the operating duties of equipment, plant or processes will assist in correctly interpreting the information on the graphs.

Further study

(a) Carbon Trust CTG077 *Monitoring and Targeting*
(b) Carbon Trust CTG075 *Degree Days for Energy Management*

A4.2 Feasibility of mitigations

The energy manager should consider any mitigations so that:

(a) it is understood which parts of the organisation are critical to business commercial success — this might be inhibited by poor implementation of energy efficiency measures;
(b) it is understood where any barriers to improvements in energy use exist — examples may include user behaviour, supply infrastructure, obsolete system controls, maintenance backlogs etc.; and
(c) they can accurately inform and influence any regular review of the business energy policy, energy strategy or associated procedures.

	Self-assessment question: Is there enough technical knowledge and understanding of how, why, and when energy loss occurred? Does the process mitigate energy loss where required?	Yes, or No?
a	There are no resources in place and no plans to mitigate energy loss.	
b	There is some knowledge and understanding of energy loss and the mitigations required. It sometimes matches the energy management team's expectations.	
c	There is technical knowledge and clear understanding of the nature of the energy loss. Feasibility processes are in place to seek mitigations. They align satisfactorily with an agreed Plan of Work, the energy policy, and energy strategy requirements.	

Mitigating energy loss needs careful consideration. Solutions to specific problems need to be part of a wider comprehensive overview of an energy management system to ensure that other areas are not affected by short term expediency.

On the demand side of the energy equation, the three main areas that any energy manager should be aware of are:

(a) improving product efficiency through deployment of better technologies and design philosophies;
(b) improving product controls through better operational philosophies and regular maintenance; and
(c) improving user behaviour so that they understand what their own responsibilities are and that energy reduction is everybody's problem.

On the supply side, the energy manager will need to consider:

(a) the resilience of supplies and back-up solutions, assuming that carbon based fossil fuel is still needed; and
(b) the reliability of renewable technology, so that fossil free (off-grid) activity is feasible.

Each demand side step will either allow energy consumption to be reduced slightly at the point of use, or less energy wastage within the distribution system. Gradually the energy equation can take on a different complexion. Introduction of energy recovery actively reduces the requirements for supply, whilst passive technology reduces the rate of energy wastage to more acceptable levels.

Having considered reduction in wasted energy from one side of the energy equation, another step is to reduce demand; this could be done in a number of ways and varying degrees of cost.

Demand side activity	Commentary
User behaviour	A campaign by the Energy Managers Association estimates that up to 40 % of wasted energy could be prevented by the behaviour change of end users. It is accepted that correctly deployed training, notices and incentives are low cost methods of obtaining reductions in energy use. Switching off appliances and equipment when not in use, provides immediate energy savings and does not require any additional technology. Behaviour change will often require a fundamental change of mindset with regards to energy consumption, especially in large enterprises. It will also require encouragement and constant input from the energy manager. It also requires the users to engage fully with the activity. There is a wealth of academic research into how best to implement behaviour change programmes.
Active technology	Active technology, especially with electrical power distribution systems, can provide automated means of saving energy. Technology such as power factor correction capacitors have been available for decades. These should be correctly sized and safe systems of work need to be in place for maintenance purposes. Harmonic filters are another technology used to improve electrical load characteristics and hence reduce wasted energy. There are also compliance issues with regards to harmonic loads on public utility supplies. The energy manager is advised to work with the operations and maintenance teams when assessing harmonics and power factors. Heat recovery technology can be deployed on mechanical ventilation systems to improve efficiency. Coupled with correctly commissioned controls this can work for both heating in winter months and cooling in summer months.

Demand side activity	Commentary
Controls	Automated controls will improve energy efficiency by reducing loads or switching off when they are not required. It is important that end users are consulted on any control system. Documents such as IEC 60364-8-1 require that any automated system on electrical loads still requires a manual override should the users need it. Incorrectly commissioned controls will not actually decrease energy consumption. Copies of commissioning test sheets should be available for the maintenance staff to monitor. Controls should be adjusted as the installation evolves or user demand changes.
Energy efficient devices	The deployment of energy efficient devices, equipment and distribution systems will reduce losses both at the point of use and throughout the engineering infrastructure. It is recommended that the energy manager liaises with the engineering staff to ensure that the correct devices are specified and installed, and that they are maintained correctly. IEC 60364-8-1 discusses the role of the client in energy efficient electrical designs including specifying the correct level of energy efficiency measures and performance, and their responsibilities in terms of the equipment deployed at the point of use. The energy manager should be a key stakeholder at this stage of the process.

Each supply side step will optimise the amount of energy required at the intake and mitigate losses at the head end of the energy distribution system.

Supply side activity	Commentary
Barycentre	A concept discussed in detail in IEC 60364-8-1, this essentially places the point of supply as close to the point of use as is possible in order to reduce transmission losses through the engineering system. This holds true for both an electrical transformer supplying a large site or a simple boiler system for space heating. The energy manager should liaise with the engineering staff to understand the configuration of the system, any likely losses and appropriate mitigations for that.
Active technology	Active technology on electrical power distribution systems will include low loss transformers. Recent European directives have increased the regulations around the acceptable levels of energy loss from transformers. Electrical transformers have a particular efficiency curve and will perform better as more load is connected. It is recommended that the energy manager liaise with engineering staff to ensure, where practical, that high efficiency transformers are used and that the connected loads allow the transformer to operate at optimum level.

Following on from Figure A1.1 the updated energy equation diagram below shows the potential effect of:

(a) reducing energy waste through passive improvements in building fabric and building services insulation;

(b) reducing inefficiency through active technology such as power factor correction and harmonic filters;

(c) reducing demand at the point of use through improved user behaviour;

(d) reducing demand by active technology such as ventilation heat recovery systems; and

(e) reducing demand by deployment of more efficient appliances and equipment.

▼ **Figure A4.1** Effects of reducing energy demand

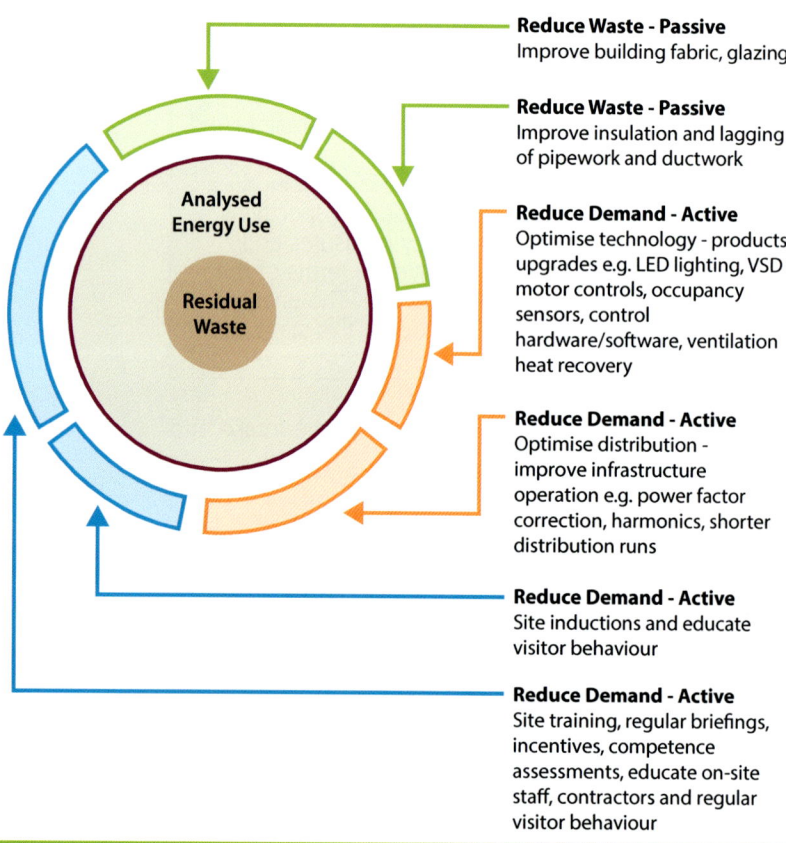

Reduce Waste - Passive
Improve building fabric, glazing

Reduce Waste - Passive
Improve insulation and lagging of pipework and ductwork

Reduce Demand - Active
Optimise technology - products upgrades e.g. LED lighting, VSD motor controls, occupancy sensors, control hardware/software, ventilation heat recovery

Reduce Demand - Active
Optimise distribution - improve infrastructure operation e.g. power factor correction, harmonics, shorter distribution runs

Reduce Demand - Active
Site inductions and educate visitor behaviour

Reduce Demand - Active
Site training, regular briefings, incentives, competence assessments, educate on-site staff, contractors and regular visitor behaviour

Analysed Energy Use

Residual Waste

Implementation example – the investigation of issues

An energy management committee, chaired by the energy manager, with staff from across the organisation , could gather anecdotal evidence of energy wastage. Backed up by automated monitoring systems, perhaps connected to the building management system (BMS), real time improvements could be made to abnormal incidents of energy loss.

Examples of outcomes are:

(a) reports to maintenance of lights that have not switched off in the day time due to failure of a photocell light sensor; and

(b) assistance in improving work flow in one part of the organisation and spotting opportunities to reduce energy consumption in the process.

A4.3 Improvements and managing change

The energy manager's role will include some involvement in improvements projects, either as the key stakeholder driving the project forward, or as an integral member of the team with overall project ownership.

Maintaining output, whilst not undermining profit margins, is the typical focus for most commercial enterprises. The capital costs involved in improvement projects to reduce avoidable energy consumption might be perceived as an unnecessary overhead and may not receive the same priority as, for example, opening a new business unit.

The energy manager must be prepared for obstacles in authorisation by the business of any energy saving project and also be aware of the methods needed for successful implementation.

Any improvements projects will require:

(a) strong support at board level within a business;

(b) a clear vision on what the project will achieve from an energy perspective;

(c) a clear vision on what the project will achieve financially in lowering ongoing energy costs;

(d) a clear understanding of the compliance implications (ESOS and similar legislation);

(e) a clear understanding of the risks of 'business as usual' in terms of energy wasted; and

(f) a clear understanding of the risks of change to lower energy consumption in terms of potentially inhibiting business growth.

	Self-assessment question: **Are there enough technical and financial resources in place to facilitate improvement and change?**	Yes, or No?
a	There are no resources in place and no plans to improve the efficiency of energy consumption or reduce carbon emissions.	
b	There are some technical and financial resources in place to facilitate improvement and change. It is low on the organisation's capital priority list and sometimes matches the energy management team's expectations.	
c	There are technical and financial resources in place to facilitate improvement and change. It is high on the organisation's capital priority list and aligns satisfactorily with the energy policy and energy strategy requirements.	

The path to successfully seeking authorisation for energy saving projects is essentially one of engagement.

The energy manager should be aware that different audiences, from end users to management to board level, will require different approaches and presentation of the facts, aspirations and realities of the proposed project.

In order to secure funding, the energy manager will need to provide a business case and be prepared to present the findings of that business case to a number of different stakeholders. Ensuring that all the facts are clearly explained is critical to the success of the business case. Some of the issues to be addressed in order to put together a successful business case are discussed further below.

Business case issues	Commentary
Basic problem analysis	Does the business case fully explain the problem with existing energy consumption and installation inefficiencies in language appropriate to the target audience? Does the business case succinctly analyse the energy management issues and why change is necessary?

Solution analysis	Does the business case succinctly describe the preferred solution and how that conclusion was reached?
	Have priorities been clearly identified and costed?
	Are the priorities dependent on other improvement projects?
	From a technical perspective, does the business case fully explain the likely outcomes and risks of doing:
	   (a) nothing — business as usual;
	   (b) something with limited resources;
	   (c) something with adequate resources; and
	   (d) something with future activities in mind?
	What energy is wasted and how will that be remedied by the preferred solution?
	How is that measured and verified?
	Are the solutions people driven, technology driven or both?
	What else can be done to improve energy performance?
	Does the preferred solution meet the requirements of the business energy policy and/or strategy?
	For any sustainable engineering solution — is the proposal appropriate to the local geography and capable of operating successfully in the local environment?

Cost implications	From a financial perspective, does the business case fully explain the implications for energy bills of doing:
	   (a) nothing — business as usual;
	   (b) something with limited resources;
	   (c) something with adequate resources; and
	   (d) something with future activities in mind?
	Does the solution offer a financial return on the capital investment that is required?
	Over what period of time is the return expected?
	Have the figures been calculated using standard calculations and communicated in easily understandable financial terms?
	Assuming the project ultimately saves ongoing energy costs, what immediate advantages are expected, such as protecting against energy price increase, or cutting energy waste?

Energy implications	Does the project impact on the business in the short term by inhibiting normal operations? Will there be any loss of services whilst the project is implemented? Is the reduction of energy consumption more important to the business ethos than reduction in basic costs? Can the project be estimated in terms of energy return on energy invested (EROEI)? The energy manager should be aware that some technology may, for instance, use three times less energy but rely on electrical energy input at three times the unit (kWh) cost. Is the use of energy within the installation transparent across the whole business through all levels to allow users clear sight of improvements and also rising trends too?

Business case issues	Commentary
Compliance implications	Is the solution part of an ongoing regulatory compliance problem or in anticipation of future changes to regulations regarding the reporting of energy use in the business? What are the implications to the business of non-compliance? What is the time frame or urgency required to meet compliance with regulations? Are the timescales for the project driven by financial cycles or legislation deadlines? Is there any dependency on technical approvals or design sign offs from external bodies?
Resource implications	Does the business case fully evaluate any additional resources required to ensure the project is successfully implemented and subsequently supported going forwards? Have the implications of the project complexity been evaluated to ensure the correct resources are in place? Has the full range of job types/roles/experience required been correctly identified? Has the proposed solution had all types of associated activities correctly evaluated and contingency is in place?
Implementation across the infrastructure	Does the business case succinctly demonstrate how implementing the energy management scheme will be delivered? What supply infrastructure improvements will there be? How will operating procedures be evaluated? What equipment improvement projects will there be and how will that be rolled out across the business?
Evaluation and review process	With whom should the results of the project be shared? How will the outcomes be compared and what benchmarks will be used? How will the improvement plan be monitored and subsequently improved? Is there a robust and continuous system in place? How does the project help with identifying and informing an improvement strategy to reduce energy consumption overall or make more efficient use of existing supply capacity? What processes will there be for a management review of the project to ensure continuous improvement?

Legislation compliance — commentary

There will be situations that provide no financial incentive to change energy performance within buildings — however new legislation may alter this situation.

An example, in the private rental sector, could be when a tenant ends up paying large utility bills as a result of poor maintenance of building fabric or building services that are owned by a landlord.

The tenant would benefit directly from any efficiencies resulting from better building insulation, or improvements in the services infrastructure, but has no incentive to invest in the structure as the landlord will reap the rewards long term. Likewise, the landlord may not invest in the building as he reaps no financial gain from doing so in the short term. This potentially leads to poor building maintenance and consequently poor performance.

These issues may be improved by the implementation of legislation, for instance, within The Energy Efficiency (Private Rented Property) (England and Wales) Regulations 2015 on minimum energy efficiency standard (MEES), due in 2018. For both domestic and commercial properties:

(a) MEES requires a minimum energy efficiency standard for a building Energy Performance Certificate (EPC) of grade E. Under the requirements of MEES, prior to being let, eligible properties should be upgraded to a minimum E EPC rating.
(b) Only appropriate, permissible and cost-effective improvements are required. Obviously, this will be tested in law.

Whilst there will be an onus on landlords to ensure that the building is compliant when tenancy agreements are made, there will also be obligations on tenants to ensure that any fit out works do not cause any detriment to an installations existing EPC rating.

Non-compliance may result in significant financial penalties being imposed under the terms of the regulations.

A time line of activities is framed within the implementation of MEES:

1   April 2016
    Residential tenants can seek landlord approval for prescribed energy efficiency improvements, and this approval must not be unreasonably refused. There are certain exemptions that may apply or the landlord can propose alternative energy efficiency measures. The expectation is that tenants will fund the improvements.

2   April 2018
    When finalising an agreement on either a new letting or a tenancy renewal, private landlords of both domestic and non-domestic property must ensure that the property has a minimum EPC rating of E.

3   April 2020 (domestic) and April 2023 (non-domestic)
    Retrospective application of the regulations to properties where a rental agreement is already in place or is occupied by a tenant.
    It will then be an offence to continue letting a sub-standard property that remains below an E rating.

Further study

(a) BRE IP2-15 *Producing the business case for investment in energy efficiency* — Dr Andy Lewry
(b) Carbon Trust CTV063 *Making the business case for carbon reduction projects*
(c) Carbon Trust CTV063 *Business case checklist*

APPENDIX B

Self-assessment questions checklist

The following series of self-assessment questions provides a supporting checklist, associated with each of the relevant sections of this Guide, to enable the reader to assess their particular installation and use the content in practice. The score values are intended to be indicative rather than adhere to existing industry evaluations.

For each self-assessment question the intent is to tick against which description most closely resembles your organisation and then allocate a corresponding score. Add all scores together and then use the scale at the bottom to give an indication of how efficient your energy management system is at present. From this an action plan can start to emerge.

It is important that the energy manager takes ownership of projects or delegates where necessary to those with more expertise in certain areas. Create a programme and set deadlines to assess any progress in terms of energy management and engineering energy efficiency.

It is recommended that the energy manager revisits the scores periodically, for instance, this might be three months for a poor score, 6 months for a better score and every 12 months for good scores. Re-evaluate the criteria and see if progress is being made.

SAQ1 Refers to Appendix A1.1 Energy management policy

	What is the Energy Policy status?	Yes, or No?	Score
a	The organisation has no Energy Policy.		
b	There is an Energy Policy, but it is incomplete, out of date or not signed off by the organisation's senior leadership.		a = 5 b = 3 c = 1
c	There is an Energy Policy in place that is accepted by the organisation's senior leadership, periodically reviewed with responsibilities delegated satisfactorily.		

SAQ2 Refers to Appendix A1.2 Energy management strategy

	What is the Energy Strategy status?	Yes, or No?	Score
a	The organisation has no Energy Strategy and reacts to business changes retrospectively. It also is not aligned to legislative requirements.		
b	There is an Energy Strategy but it lacks resources and active management. It does not always cope with future plans or changes to the organisation's needs.		a = 5 b = 3 c = 1
c	The Energy Strategy aligns satisfactorily with the Energy Policy and legislative requirements. It is reviewed and able to cope in advance of any changes to the organisation.		

SAQ3 Refers to Appendix A1.3 Energy management procedures

	What is the Energy Procedures status?	Yes, or No?	Score
a	The organisation has no Energy Procedures of its own and does not use recognised standards or industry codes to manage energy at an operational level. It also is not aligned to legislative requirements.		
b	There are some Energy Procedures but they are not complete or lack resources and active management. The procedures do not match the organisation's needs.		a = 5 b = 3 c = 1
c	The Energy Procedures align satisfactorily with the Energy Policy, Energy Strategy and legislative requirements. They are reviewed and able to adapt to any changes to the organisation.		

SAQ4 Refers to Appendix A1.4 Engineering design for energy management

	What is the interface to engineering design? Has energy management been considered for new systems, new builds, or retro fit projects?	Yes, or No?	Score
a	There is no direct involvement by the organisation's energy management team as stakeholders in the design of new engineering systems.		
b	There is some consultation of the organisation's energy management team but it is inconsistent. The process does not always match the organisation's needs or the energy management team's expectations.		a = 5 b = 3 c = 1
c	The engineering design process aligns satisfactorily with the Energy Policy, Energy Strategy and legislative requirements. It is reviewed and projects take into account the existing infrastructure and perceived energy risk assessments.		

SAQ5 Refers to Appendix A2.1 Procurement and energy

	What is the Energy Procurement Plan?	Yes, or No?	Score
a	Energy supplies are purchased directly from the energy suppliers and user equipment is purchased as and when required.		
b	There is some consultation of the organisation's energy management team but it is inconsistent. The process does not always match the organisation's needs or the energy management team's expectations.		a = 5 b = 3 c = 1
c	The energy procurement process aligns satisfactorily with the Energy Policy and Energy Strategy requirements. It optimises the available tariffs and the use of energy efficient equipment.		

SAQ6 Refers to Appendix A2.2 Roles and responsibilities

	Are energy management roles and responsibilities defined?	Yes, or No?	Score
a	There is no clear structure on energy management and the bill payer bears sole responsibility for energy use.		
b	There is some understanding across the organisation of individual roles. It is ad hoc and driven by individuals and does not necessarily match the energy management team's expectations.		a = 5 b = 3 c = 1
c	Energy management roles and responsibilities are clearly defined and properly delegated. They align satisfactorily with an agreed Plan of Work, the Energy Policy, and Energy Strategy requirements.		

SAQ7 Refers to Appendix A2.3 User behaviour

	Is user behaviour and associated patterns of use clearly understood?	Yes, or No?	Score
a	There is no clear understanding or influence over user behaviour to avoid energy wastage.		
b	There is some understanding across the organisation of user behaviour. It is driven by individuals and sometimes matches the energy management team's expectations.		a = 5 b = 3 c = 1
c	User behaviour is clearly understood and programmes to communicate and improve are in place. They align satisfactorily with an agreed Plan of Work, the Energy Policy, and Energy Strategy requirements.		

SAQ8 Refers to Appendix A2.4 Maintenance

	Are communication channels in place between energy managers and operational maintenance teams?	Yes, or No?	Score
a	There is no link up between energy managers and maintenance teams. Priorities for energy efficiency in design and operation are not coordinated.		
b	There is some understanding across the organisation of the link between maintenance and energy management. The communication channels are ad-hoc and driven by individuals rather than strategy. It sometimes matches the energy management team's expectations.		a = 5 b = 3 c = 1
c	The link between maintenance and energy management is clearly understood and programmes to communicate and improve are in place. They align satisfactorily with an agreed Plan of Work, the Energy Policy, and Energy Strategy requirements.		

SAQ9 Refers to Appendix A3.1 Performance audits

	Is there a process in place for systemic evaluation and auditing of the energy consumption?	Yes, or No?	Score
a	There is no auditing in place beyond that of checking the energy bills.		
b	There is some understanding of the need to periodically check energy consumption. It is driven by individuals rather than strategy. It sometimes matches the energy management team's expectations.		a = 5 b = 3 c = 1
c	There is clear understanding of the need to regularly check energy performance at a local level. Processes are in place. They align satisfactorily with an agreed Plan of Work, the Energy Policy, and Energy Strategy requirements.		

SAQ10 Refers to Appendix A3.2 Benchmarks

	Does the energy management team understand what benchmarks are in place? Are they reviewed for changes of use to buildings?	Yes, or No?	Score
a	There are no existing benchmarks in place to meaningfully evaluate the current performance of the building.		
b	There are some benchmarks in place but they have not been reviewed or updated. Building performance sometimes matches the energy management team's expectations.		a = 5 b = 3 c = 1
c	There is clear understanding of the commissioned building performance benchmarks. These are regularly checked and updated where appropriate. Processes are in place. They align satisfactorily with the Plan of Work, the Energy Policy, and Energy Strategy requirements.		

SAQ11 Refers to Appendix A3.3 Analysing losses

	Is there a process in place for systemic evaluation losses in the energy consumption? Is there a regular review of the installation?	Yes, or No?	Score
a	There is no understanding of the losses nor available skills to assess them.		
b	There is some understanding of how to assess the energy losses in the installation. It sometimes matches the energy management team's expectations.		a = 5 b = 3 c = 1
c	There is clear understanding of how to regularly check the installation for energy performance losses at a local level. Processes are in place. They align satisfactorily with an agreed Plan of Work, the Energy Policy, and Energy Strategy requirements.		

SAQ12 Refers to Appendix A4.1 Review of targets

	Is there a process in place for systemic evaluation of energy consumption targets? Does the process include adjustments where required?	Yes, or No?	Score
a	There are no energy targets in place.		
b	There is some understanding of energy targets for the installation and how to monitor them. It sometimes matches the energy management team's expectations.		a = 5 b = 3 c = 1
c	There is clear understanding of how to evaluate energy targets. Monitoring technology and processes are in place. They align satisfactorily with an agreed Plan of Work, the Energy Policy, and Energy Strategy requirements.		

SAQ13 Refers to Appendix A4.2 Feasibility of mitigations

	Is there enough technical knowledge and understanding of how, why, and when energy loss occurred? Does the process mitigate energy loss where required?	Yes, or No?	Score
a	There are no resources in place and no plans to mitigate energy loss.		
b	There is some knowledge and understanding of energy loss and the mitigations required. It sometimes matches the energy management team's expectations.		a = 5 b = 3 c = 1
c	There is technical knowledge and clear understanding of the nature of energy loss. Feasibility processes are in place to seek mitigations. They align satisfactorily with an agreed Plan of Work, the Energy Policy, and Energy Strategy requirements.		

SAQ14 Refers to Appendix A4.3
Improvements and managing change

	Are there enough technical and financial resources in place to facilitate improvement and change?	Yes, or No?	Score
a	There are no resources in place and no plans to improve the efficiency of energy consumption or reduce carbon emissions.		
b	There are some technical and financial resources in place to facilitate improvement and change. It is low on the organisation's capital priority list and sometimes matches the energy management team's expectations.		a = 5 b = 3 c = 1
C	There are technical and financial resources in place to facilitate improvement and change. It is high on the organisation's capital priority list and aligns satisfactorily with the Energy Policy and Energy Strategy requirements.		

Item	Self-assessment question	Score
1	What is the Energy Policy status?	
2	What is the Energy Strategy status?	
3	What is the Energy Procedures status?	
4	What is the interface to engineering design?	
5	What is the energy procurement plan?	
6	Are energy management roles and responsibilities defined?	
7	Is user behaviour and associated patterns of use clearly understood?	
8	Are communication channels in place between energy managers and operational maintenance teams?	
9	Is there a process in place for systemic evaluation and auditing of the energy consumption?	
10	Does the energy management team understand what benchmarks are in place? Are they reviewed for changes of use to buildings?	
11	Is there a process in place for systemic evaluation losses in the energy consumption? Is there a regular review of the installation?	
12	Is there a process in place for systemic evaluation of energy consumption targets? Does the process include adjustments where required?	
13	Is there enough technical knowledge and understanding of how, why, and when energy loss occurred? Does the process mitigate energy loss where required?	
14	Are there enough technical and financial resources in place to facilitate improvement and change?	
	Total	

How does your organisation's score compare?

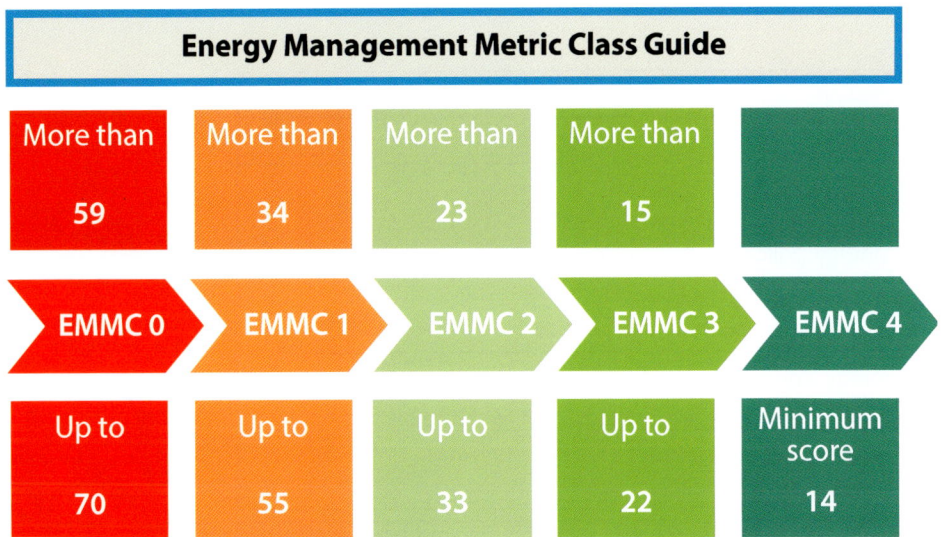

Energy Management Metric Class Guide

More than 59	More than 34	More than 23	More than 15	
EMMC 0	EMMC 1	EMMC 2	EMMC 3	EMMC 4
Up to 70	Up to 55	Up to 33	Up to 22	Minimum score 14

Summary of energy management metric class (EMMC)

EMMC 0	Very poor energy management system
EMMC 1	Poor energy management system
EMMC 2	Reference (or benchmark) energy management system
EMMC 3	Advanced energy management system
EMMC 4	Optimised (or exemplar) energy management system

Supplementary energy management questions for self-assessment

The following table provides additional questions that the energy manager may wish to evaluate as they build a robust energy management system for their particular organisation, or in support of a business case to advocate for improvement projects or additional resources.

The table is provided as a guide only. There may be additional questions to be added to this table by the energy manager to suit their particular role, organisation or the estate being managed.

Item	Supplementary question	Self-assessment comments
SAQ1	Energy policy	Purpose
1.1	Does the energy manager have board level support and/or an assigned champion for the energy policy?	
1.2	Does the energy policy seek to reduce energy consumption?	
1.3	Are sustainable energy sources part of the forward plan?	
1.4	Have key stakeholders throughout the organisation all signed off on the policy?	
SAQ2	Energy strategy	Purpose
2.1	Do you understand existing energy requirements within all costs centres?	
2.2	Has the energy strategy been reviewed to account for future projects/installations?	
2.3	Does the energy strategy align with the overall business strategy and vice versa?	
2.4	Does the energy strategy highlight risks to site capacity and resilience of supply?	
2.5	Does the energy strategy account for business priorities?	
2.6	Does the energy strategy account for all known barriers to implementation?	
2.7	Does the energy strategy specify priorities for reducing losses, inefficiency and waste?	
SAQ3	Energy procedures	Purpose
3.1	Do the energy management procedures correctly assess risk in terms of engineering systems and user behaviour?	
3.2	Are the energy management procedures regularly reviewed?	
3.3	How are changes communicated to relevant stakeholders and energy users?	
3.4	Do the energy procedures clearly set out how energy consumption will be monitored?	
3.5	Do the energy procedures clearly set out how energy performance will be reviewed?	

SAQ4	Engineering design interface	Purpose
4.1	Is there liaison between energy manager and engineering design manager?	
4.2	Do new additions, designs, refurbishments meet the site strategy?	
4.3	Is the energy source located so that the infrastructure works efficiently?	
4.4	Are appropriate meters in place with a clearly defined metering strategy?	
4.5	Does the design consider whole life cycle energy performance and associated costs?	
4.6	Have energy efficient appliances, equipment and controls been specified?	
4.7	Have the practicalities of building, maintaining and disposing been considered?	

SAQ5	Energy procurement plan	Principles
5.1	Do the grid-connected supplies have sufficient capacity for current?	
5.2	Can future needs be accommodated within the existing capacity or does the grid supply need upgrading?	
5.3	Are the energy contract tariffs correct, and appropriate for existing patterns of use?	
5.4	Is a review required of the energy contract and is the contract up for renewal?	
5.5	Have recent bills been checked? Are they clear and understood? Do they need to be challenged?	
5.6	Is there a plan for seasonal use of energy? Additional fuel supplies for oil tanks, etc.?	
5.7	Is there a plan for alternative supplies if local (CHP) generation is not available for maintenance purposes?	

SAQ6	Roles and responsibilities	Principles
6.1	Is the role of the energy manager clearly defined by the organisation?	
6.2	Are the roles of all other key stakeholders in the energy management system clearly defined?	
6.3	Has all appropriate training been identified and other channels in place for knowledge dissemination of energy management issues?	
6.4	Are there delegated responsibilities to manage energy during the proactive stages?	
6.5	Are there delegated responsibilities to manage energy during the reactive stages?	

SAQ7	User behaviour and patterns of use	Principles
7.1	Do site induction processes include energy management briefings and is it clear on what is expected of users?	
7.2	Is it clear to users the impact their consumption of energy may have on the performance of the installation?	
7.3	Are there appropriate processes in place to provide feedback from users on unnecessary energy consumption?	
7.4	Does the energy management system provide the appropriate levels of influence on user behaviour?	
7.5	Does the energy management system provide the appropriate levels of training for users and operators?	

SAQ8	Engineering maintenance interface	Principles
8.1	Is there liaison between energy manager and engineering maintenance manager?	
8.2	Has the maintenance team been adequately trained on new additions and refurbishments to the wider estate?	
8.3	Are functional checks carried out to ensure energy efficient appliances and controls are operating satisfactorily?	
8.4	Is there a schedule available for the specified energy efficient appliances, equipment and controls?	
8.5	Have the practicalities of maintaining and disposing been considered?	

SAQ9	Auditing of energy consumption	Parameters
9.1	Is the metering and sub-metering strategy for the installation understood for each energy type and implemented correctly?	
9.2	Are the meters checked periodically and the energy consumption values recorded?	
9.3	Is there a continuous record of meter readings to enable analysis of consumption trends to take place?	
9.4	Have appropriate values from the electrical meters been noted to allow power quality analysis (check operational efficiency)?	
9.5	Is there sufficient information to enable various load types to be analysed separately?	
9.6	Does the audit team require competence checks and additional training?	

SAQ10	Review of energy benchmarks	Parameters
10.1	Are the design benchmarks available?	
10.2	Has there been a change of use in processes, work flows or occupancy levels? If yes, should the benchmarks be updated?	
10.3	Have there been recent additions to the installation or refurbishments? If yes, should the benchmarks be updated?	
10.4	Are there resources available to analyse the data sets or do external consultants need to be retained?	
10.5	Does the energy management team have access to local weather data?	

SAQ11	Evaluation of energy losses	Parameters
11.1	Have losses through the building fabric been investigated?	
11.2	Have losses, or additional consumption, through the engineering infrastructure and appliances been investigated?	
11.3	Have losses, or additional consumption, associated with user activities been investigated?	
11.4	Is the process of monitoring and targeting properly understood?	
11.5	Is there a requirement for a measurement and verification plan?	

SAQ12	Evaluation of energy targets	Progression
12.1	Have energy management targets been set?	
12.2	Have users been consulted and are the targets realistic?	
12.3	How are targets communicated to users and what updates or feedback do they get?	
12.4	Is it clear what methods have been used for energy management forecasts and the likely range of results and accuracy?	
12.5	How often are the targets reviewed and updated?	

SAQ13	Mitigation of energy loss	Progression
13.1	Is investment required to improve infrastructure efficiency and engineering design?	
13.2	Is there a need to replace appliances and equipment? Is the existing equipment due for replacement anyway?	
13.3	Can energy controls philosophy be improved? Does the maintenance regime need closer management?	
13.4	Have user behaviour issues been identified that can be improved on, either by encouragement or regulation?	
13.5	Is it clear what passive improvements can be made to the installation fabric?	
13.6	Is it clear what active improvements can be made to the installation engineering services?	

SAQ14	Improvement and change	Progression
14.1	Is there a clear process for instigating a business case for improvements?	
14.2	Does the proposal fully explain the problem, the analysis and the preferred solution? Are there other options too?	
14.3	Has an adequate risk analysis been completed to support any proposal?	
14.4	Has a board level champion been identified to support the business case?	
14.5	Are there legislative or compliance implications if the improvements are not carried out?	
14.6	Have the impacts on the business been properly evaluated – short, medium and long term?	
14.7	Are there resource implications to ensure success of the preferred solution?	
14.8	Does the preferred solution match the existing energy policy and strategy, or will these need to be updated too?	

In addition, the following Carbon Trust document also provide checklists that can be downloaded from their website and used to assist in the evaluation of various parts of the energy management process.

(a) CTG 054 *Energy Management*
- **i** Energy Management Matrix
- **ii** Energy Management Assessment

(b) CTG 055 *Energy Surveys*
(c) CTV 067 *Making the business case for a carbon reduction project*

Overview of technical and engineering considerations

The following section is a guide to some of the technologies that will be considered when improving the energy efficiency of an installation. The purpose is to provide energy managers with a background overview to understand some of the considerations they will need to take into account.

Energy managers should understand that further discussion will be necessary with the appropriate designers, architects, engineers and specialist contractors to fully utilise the efficiencies that may be on offer for any particular installation.

C.1 Building fabric

Assessment of the building envelope often represents the first step to any energy management programme, especially in older, existing buildings that are not built to current standards. One of the biggest concerns is that buildings that are not designed correctly will leak heat through the fabric of the building and through open doors and windows.

Criteria		Commentary
Preventing fabric heat loss	Why?	Installation efficiency improves if the building fabric is correctly insulated to mitigate the effects of heat loss. Glazing, walls, ceilings and floors should all be assessed.
	How?	Suitable modern designs for new buildings to relevant building codes such as the UK Building Regulations will ensure the correct levels of insulation and other measures to reduce heat loss. Existing buildings can be improved during any refurbishment programmes. Alternatively, existing buildings can be surveyed by competent personnel and assessments made on retrofitting additional insulation where appropriate.
Integration of photovoltaic panels	Why?	Putting the external surface of the building to good use to provide local generation of electricity and offset grid supplies.
	How?	Provided in new builds by a better architectural solution to solar generation of energy using integrated photovoltaic panels on the roof and also in the walls on buildings that have appropriate orientation to the sun's path.

Criteria		Commentary
Pros and cons of airtightness	Why?	Allowing air that has been heated to escape from a building through open windows, or doors is wasteful. Equally the colder external air can infiltrate through these openings and other apertures as draughts, which is uncomfortable for occupants and places more demands on the heating system.
	How?	Carefully controlling the passage of heated air means the internal temperature is better controlled, requiring less energy to maintain a comfortable and satisfactory environment. However, indoor air quality is also of major concern. Unless there is a regular supply of fresh air throughout the whole day, then the internal environment within a building will become stale and unhealthy to the occupants. Metrics such as air changes per hour should be used to ensure a healthy environment. Using heat recovery systems is an efficient way to recover heat from stale air and allow it to temper the incoming fresh air to a warmer temperature. This reduces the additional energy needed to maintain a satisfactory temperature.
Thermal bridge	Why?	This allows heat to escape and also allows the conditions for condensation in the internal spaces to exist. In a double skinned building, for instance, with a cavity wall, it is often caused by poor workmanship allowing cement to bridge the gap between the two brick skins of the building. Often identified by damp patches on internal walls that will not dry out or accept chemical treatment processes.
	How?	Resolving this will need specialist advice and assistance from qualified builders and specialists. Small remote cameras may be needed. Parts of the existing brick wall may need to be removed to allow remedial works to take place.
Daylight	Why?	There will be less use of artificial lighting internally if there is more use of glazing to allow natural light to penetrate the buildings more comprehensively. However, in older buildings that still retain single glazing panels there would be considerable heat loss.
	How?	Use of daylight sensors on the artificial lighting scheme will allow lighting to either dim or switch off when there is sufficient daylight to allow tasks to take place.

C.2 Metering, monitoring and targeting

Assessing energy consumption through the use of just the main fiscal energy meters is not necessarily enough to actively manage energy, especially on larger installations with multiple departments and buildings. There will be multiple uses of equipment, areas and the occupancy levels will vary throughout the day.

In order to effectively monitor energy use sub-metering is preferred. For large sites with multiple buildings it is advantageous to provide each building with intake sub-meters for each utility service. From there energy consumption can be measured periodically, continuously monitored and compared for trends in use. Targets should be set to reduce consumption over a period of time.

Publications from various organisations provide advice on what type of meters to place where and how to interpret the data correctly. It is important to have a metering strategy and careful planning should be undertaken to facilitate this. Up to date schematics of services within buildings will assist in placing sub-meters on loads that have the greatest impact on energy consumption and where closer energy management will have the best results.

CIBSE Guide TM39 will assist with design strategy and the process of identifying the best location for sub-meters. IET Standards *Guide to Metering Systems* will assist with more practical considerations of specification, installation and use.

Accurate meter readings will also assist with compliance obligations.

Increasing numbers of installations will also be affected by forthcoming changes within the electricity market on P272 and half-hourly meters.

The water supply market will also be changing with metering being central to the ability of businesses being able to change their water supplier.

Further study

(a) IET Standards *Guide to Metering Systems – Specification, Installation and Use*
(b) Carbon Trust (CTV027) – *Metering technology overview*
(c) Carbon Trust (CTG008) – *Monitoring and targeting – in depth management guide*
(d) CIBSE TM39 – *Building Energy Metering*

C.3 Active technology and control systems

Optimising the control of an engineering system will reduce the consumption of energy. This is especially true when an area is not occupied.

Implementation and performance — building management

Building controls, whether stand-alone units or full building energy management systems (BEMS), are designed to provide a comfortable climate for building occupants while ensuring this is delivered with the lowest possible energy consumption.

Controls can be used to manage heating systems, cooling systems, air conditioning systems, lighting systems and blinds, as well as fire and security systems and lifts.

They can also be used to directly collect and display data from meters. Energy data can then be displayed on the BEMS; having good quality data about actual energy consumption is the key to achieving an energy efficient building.

Demand-based control is the most energy efficient approach; turning systems off when not needed or, if this cannot be done, then at least turning them down.

To operate effectively, controls need the appropriate functionality and should be capable of two-way communication with the building operators to ensure that operational data is collected and any issues are highlighted immediately.

The performance of controls can be assessed by BS EN 15232 *Energy performance of buildings — Impact of building automation, controls and building management.* This document has a series of classes describing the energy performance.

Optimisation — building management system (BMS)

BMS optimisation can be a relatively quick and cost effective way to enhance the energy performance of an existing building. Traditionally, control of a building may have been little more than a thermostat with a timer. Modern BMS, though, can be a complex network of controls and sensors that make it a more sensitive, adaptable and demand-based system.

Contemporary BMS systems can achieve huge energy performance gains without compromising building comfort. These changes can often be made by on-site or remote software changes, negating the need for large capital expenditure on new plant.

The BMS can be configured to react to such things as:

(a) internal and external ambient temperature and air quality;
(b) internal and external lighting conditions;
(c) levels and number of occupants;
(d) times of occupancy; and
(e) exception programming, for example, bank holidays and other similar events.

Controls — considerations

When employing optimisation, night purge or free cooling, specialist assistance is advisable to address considerations such as:

(a) specific outside climate conditions — humidity and temperature could adversely affect internal environments;
(b) out of hours occupancy — overrides may be needed if users are in the building before or after normal working hours;
(c) security — automated opening of windows could compromise overnight security;
(d) air quality — risk of air borne contaminants introducing foul air particularly in sterile environments — filters mitigate this, but reduce the energy efficiency of natural ventilation; and
(e) maintenance — coordination needed if controls need overriding to enable a safe system of work for maintenance purposes.

Criteria		Commentary
Optimisation of controls philosophies	Why?	Incorrect commissioning or neglected controls systems following a change of use does not allow a heating or hot water system to keep pace with demand.
	How?	Robust control philosophies for heating and hot water that are regularly reviewed and updated as the use of a building develops over time will mitigate potential losses.
Optimisation of energy performance	Why?	Initial settings may not always be appropriate for an installation; equipment may fail and external influences from the weather may increase demand. Energy may be wasted by end users in a variety of ways through simple mistakes or neglect.
	How?	Ensure adequate metering and monitoring. Typical metrics for evaluating energy performance should be made available and compared to the meter outputs on a regular basis.

Controls — sensor types

Sensors are deployed in engineering systems to provide automated basic or complex control over a wide range of conditions and operations. There are numerous types of sensor and the more common ones include:

(a) temperature sensors — may be room, outside or duct mounted to measure and control internal conditions and external temperatures;

(b) occupancy sensors — presence or absence detection of occupants to control lighting and HVAC; can be passive infra-red (PIR), microwave, ultrasonic;

(c) lux sensors — measures ambient natural daylight to either enable or disable artificial lighting — can also be used for zonal control where differing levels of light are present or required;

(d) air quality sensors — uses include measuring levels of carbon dioxide, carbon monoxide and other odours or contaminants, normally linked to a fan/variable speed drive to control speed and thus energy;

(e) humidity sensors — measures relative humidity to ensure correct operational conditions and comfort levels — could have an energy benefit as thermal transmittance is adversely affected by high humidity; and

(f) pressure and velocity sensors — either for fluid or air filter to and optimises control of valves, dampers, fans, blowers, flaps and pumps that in turn can lead to energy reductions.

Other measures include the use of variable speed drives on motors and thermostatic radiator valves.

▼ **Table C3.1** Summary of control types and their function

Control	Uses and applications
Building management system (BMS)	(a) A BMS is a computer-based system which integrates building functions, i.e. heating ventilation and air conditioning (HVAC), fire, security, power systems and lighting. (b) Available in pre-programmed or programmable formats. (c) Systems are available for all types of businesses and sizes of buildings.
Building energy management system (BEMS)	(a) BEMS control and monitor plant such as lighting and HVAC in order to specifically address energy use. (b) BEMS does not integrate all parts of the building as a BMS does, i.e. control of security and fire protection systems is not normally included.
Demand control or zone control	(a) Demand control enables the HVAC system to operate until the demand is satisfied, e.g. cooling, hot water, radiators and air handling. (b) Demand control can be linked in to CO_2 sensors or footfall sensors. (c) Floor control dampers can also help reduce the load on an air handling unit (AHU), when there is no demand on the floor or the floor can be isolated if unoccupied in multi-let premises. (d) An example of a suitable application is a commercial property with more than one HVAC system or piece of plant. An existing BMS/BEMS will need to be in place.
Sequencing	(a) Sequencing can be a stand-alone control or via the BMS/BEMS. (b) Controlling the number of boilers required to meet the current heating load of the building. (c) This increases plant service life and allows for maintenance down-time during periods of low load.
Weather compensation	(a) Controlling the indoor temperature of the building independently of increases or decrease in outdoor temperature. (b) Enables energy savings to be achieved by reducing the heating system's operating (flow) temperature. (c) Can also be known as variable temperature control. (d) Weather compensation can be provided as part of the BMS/BEMS or as a stand-alone control. (e) If humidity control is a requirement, e.g. in swimming pools, art galleries, specific humidity controls can be deployed.
Boiler load optimisation	(a) Stand-alone control which prevents boilers from dry cycling and reduces energy costs. (b) Flow and return sensors monitor the temperature of each boiler. (c) Complements existing controls and can be integrated. (d) Installed to each boiler. (e) Boiler optimisation can be programmed as part of the BEMS and some have a standard strategy package to do this.
Optimum start/stop	(a) A time schedule should be set up to control plant and equipment to fit in with the occupancy of times of a building. (b) The time schedule will also be used to provide optimum start (and stop) of the HVAC plant to ensure comfort conditions are achieved for the start and finish of occupancy.

Control	Uses and applications
Occupancy controls	(a) Mainly used in lighting systems, though they can also be used for fast-response extract fan systems in bathroom areas. (b) Heating and cooling systems tend to be too slow in their response to be effectively controlled by occupancy sensors. (c) However, actual occupancy sensors can be used on meeting rooms and cellular offices to control fan coil unit (FCU) fans and/or air conditioning (AC) units. The control strategy would be engineered to hold the room at a setback temperature of say 16 °C or 18 °C and will bring the room to comfort conditions when occupancy is detected. (d) There are typically four types of sensors: passive infrared (PIR) sensors, ultrasonic sensors, microwave sensors and audio sensors.
Variable controls	(a) Controlling the speed of drives and fans when full speed is not required will deliver cost savings. (b) When designing a new installation, it is more energy efficient to use two port valves rather than traditional three port valves. This allows a pressure sensor to be used to reduce the speed of the pumps as valves close and system pressure increases. (c) Variable controls can ensure changes in production and/or occupancy are fully optimised.
Interlock controls	(a) Prevents unnecessary energy use and plant operation. For example, if doors or windows are opened, the interlock controls prevent the boiler(s) or air conditioning from operating.

Controls — the process

(a) Understand what controls that the installation already has.
(b) Determine the business needs.
(c) Determine the functionality required of the controls.
(d) Select an appropriate servicing strategy.
(e) Match these against a class of BS EN 15232.
(f) Ensure the chosen class has the required functionality.
(g) Produce a comprehensive specification.

Controls — caution

(a) The following key issues should be noted and addressed:
   i   specification breaking — procurement routes and 'value engineering';
   ii   occupancy patterns — schedules and density;
   iii   future proofing — flexibility and upgrades;
   iv   links to monitoring and targeting (M&T) — optimisation systems;
   v   verification/certification;
   vi   commissioning — initial set-up and an on-going process;
   vii   training;
   viii   maintenance requirements — planned upgrades;
   ix   management reporting; and
   x   additional functionality — critical services/alarms.

(b) It is strongly advised that expert advice is sought at the stage where internal capabilities are exceeded; controls are not something to learn along the way.

References

Some of the text within this section has been adapted from the following original sources by kind permission of Dr Andy Lewry at BRE:

(a) *Understanding the choices for building controls.* BRE IP 1/14. Bracknell, IHS BRE Press, 2014. Joint publication with ESTA and sponsored by Siemens.
(b) *Energy Manager's Guide to Building Controls,* Energy Manager's Association.
(c) BRE/ESTA Briefing Paper: *Energy management and building controls.*

C.4 Electrical power systems and equipment

The operational performance of an energy efficient electrical installation can be constrained by inefficiencies in the design of the electrical infrastructure.

For existing installations, the legacy of past projects may be difficult to overcome, but with careful analysis some improvements should be made. It is important that refurbishments do not compound the mistakes of the past, but instead take the chance to provide real energy efficiency changes.

IEC 60364-8-1:2014 draws the electrical designer's attention to various criteria:

Criteria for infrastructure		Commentary
Location of the main substation	Why?	Installation efficiency improves if loads are closer to the point of supply. Reducing the distance between load and supply mitigates losses in the electrical distribution infrastructure.
	How?	IEC 60364-8-1:2014 demonstrates a calculation method for barycentres to ensure the distance to the respective load centre locations is minimised with respect to the principal points of supply.
Optimisation of the transformer	Why?	Transformers are reasonably efficient, especially when the load is near the transformer rating. However, there are still improvements in transformer technology that can be optimised.
	How?	New transformer technology reduces the residual iron losses and variable copper losses within the transformer windings. Manufacturers' data should enable an analysis of these losses.
Optimisation of the wiring system	Why?	The design of the wiring system is critical to ensuring the efficiency of the whole electrical infrastructure.
	How?	Barycentre concepts for the location of the supply. Mitigation of wiring losses by appropriate designs for voltage drops, undersized cables, power factors and harmonics. Robust control philosophies around which loads are grouped together, what external parameters there are, and the associated controls and measurements.

Criteria for infrastructure		Commentary
Optimisation of power factors	Why?	Power factors of large equipment will make an electrical infrastructure less efficient. Poor power factors can leave installations exposed to financial penalties form the supply companies.
	How?	Correct selection of equipment and the installation of compensatory capacitors reduce the impact of power factors on the wiring system. The concepts around power factor measurement and correction are understood by those operating in the electrical engineering field. These are usually placed at the main switchboard of an installation. Local power factor correction should also be considered for large items of equipment that are some distance from the switchboard, for example air handling units, chillers, lifts and process plant.
Reduce the effects of harmonics	Why?	Harmonics waste energy but can also damage electronic equipment connected to the installation. Harmonics can impact upon the external supply network. Penalties can be applied to installations that do not control harmonics.
	How?	Electrical designs should select the correct equipment. Other mitigations are harmonic filters, correctly sized cables and distribution philosophies that decouple harmonic loads from linear loads.
Measure energy and power (kWh and kW)	Why?	Measuring energy usage (kWh) and power (kW) is an important component of energy management.
	How?	For smaller installations one meter at the main distribution board may be sufficient. For new, larger installations there is a need for additional information using local meters on distribution boards and occasionally large loads (mechanical services).

Criteria for infrastructure		Commentary
Measure voltage levels	Why?	The further a load is from the origin of a building's supply the lower the voltage will be at the point of use.
		Technologies exist that suggest there is scope to reduce energy waste by reducing voltage, particularly in older installations with legacy electrical equipment. It should be noted that the effect and benefits of lower operating voltages varies depending on the connected type of electrical load. Voltage reduction will only yield energy savings in resistive loads which are uncontrolled or not regulated. It should also be recognised that lowering the voltage causes an inevitable reduction in useful output. It should be considered only where this can be tolerated and examples may include filament lamps and toilet extract fans.
		Electrical equipment, with a defined amount of output, will continue to draw substantially the same amount of input electrical energy, either by running for longer at reduced power or by drawing higher current to deliver constant power. In the latter case energy input may well increase, not decrease, because of higher resistive losses. Intermittent thermal loads such as ovens, kettles and laser-printer platens will always use more energy at lower voltage — this is necessary to balance the increased heat loss per cycle consequent on extended cycle times.
		Retrofit voltage reduction equipment consists of transformers. These also incur standing losses all the time they are energised. Equipment suppliers are very unlikely to mention this additional overhead.
	How?	Measuring the voltage will assist with diagnostic assessment and instigating remedial actions.

Further study

(a) IEC 60364-8-1 *Low Voltage Electrical Installations: Energy Efficiency*

(b) Carbon Trust CTG045 *Voltage Management An introduction to technology and techniques*

(c) IET *Designers Guide to Energy Efficient Electrical Installations*

C.5 Lighting

Criteria for lighting loads		Commentary
	Why?	Installation efficiency improves if the luminaires are located correctly, with the correct levels of illumination (including task and ambient) and also the optimum automatic and manual controls. Energy efficient technologies such as LED lighting may improve lighting loads but caution should be used on colour rendition and performance depending on the application.
	How?	Lighting design philosophies using lower general ambient light and higher task lighting may be more energy efficient. Traditional designs use uniform levels of light across larger areas, typically with higher lighting levels and hence more energy required. International lighting standards provide appropriate lighting design levels for various applications including particular task or general areas. Building Regulations indicate the required energy performance of lighting and the use of appropriate calculations to factor in the controls used. Evaluation of energy efficient lighting can be expressed as: (a) watts per square metre (W/m^2); or (b) watts per square metre per 100 Lux ($Wm^2/100Lux$). These show how much energy may be used but they do not take into account controls, automatic switching or the period of time during which it may be consumed. Approved Document L of the UK Building Regulations (2016) allows use of the Lighting Energy Numeric Indicator (LENI) to express energy per square metre per year ($kWh/m^2/yr$). This assesses the lighting load and its losses, automatic controls, maintenance and lifecycle performance.

Further study

(a) BS EN 12464-1 *Light and lighting. Lighting of work places. Indoor work places.*
(b) BS EN 12464-2 *Light and lighting. Lighting of work places. Outdoor work places.*
(c) IET Standards *Code of Practice for the Application of LED Lighting Systems.*
(d) IET Standards *Recommendations for Energy Efficient Exterior Lighting Systems.*
(e) CIBSE and the Society of Light and Lighting for guidance on the design of lighting installations, the implementation of controls and the evaluation of energy.

C.6 Heating and hot water

The design, installation, commissioning, operation and maintenance of the heating and hot water systems are critical to successful outcomes for an energy manager. Heating and hot water are often coupled in common systems and drive their heat from similar appliances. Pipework often uses the same routes. Energy losses, especially in older installations, are often encountered and need to be addressed as a priority.

Criteria for infrastructure		Commentary
Location of the main heating boilers	Why?	Installation efficiency improves if loads are closer to the point of supply. Reducing the distance between load and supply mitigates losses in the heating and hot water distribution infrastructure.
	How?	Barycentres concepts could be used to ensure the distance to the respective load centre locations is minimised with respect to the principal points of supply. Matching the correct boilers for loads and using district heating for larger estate wide installations can provide scaled up efficiencies. Modular boiler systems are better at matching varying demands on larger heating systems.
Optimisation of distribution pipework	Why?	Heating and hot water pipework that are not insulated or are too long can lose energy.
	How?	Pipework insulation throughout the pipework infrastructure to mitigate losses and ensure as much of the required heat is actually delivered to the point of use when the consumer requires it. Shorter runs from the heat source to the point of use helps to reduce wasted water when waiting for the required temperature.
Hot water storage	Why?	Setting a lower temperature for space heating, if it is tolerable for the occupants, could be a useful strategy to save energy. However, using a similar strategy for hot water should be used with extreme caution.
	How?	There are clear guidelines on the correct temperature setting for hot water and associated storage tanks to prevent the spread of water borne diseases such as legionella and pseudomonas. Careful controls via thermostats need to be used to define the correct set points above and below the temperature zone at which such problems proliferate.

Energy loss of boilers in use — case studies

The following demonstrates some of the issues surrounding correct installation, proper commissioning and satisfactory system controls.

Dry cycling
This occurs when a boiler loses heat to its surroundings, known as standing losses. Water temperature in the boiler falls below the existing set-point of the boiler thermostat.
The dry cycle is the action of the boiler energising to restore the temperature of the water within the boiler. This wastes energy because there is no actual demand for heat at that time from the building.
Boiler dry cycling has significant potential for energy loss through incorrect set up of the installation and it is important that controls are in place to prevent it. Performance and settings should be reviewed regularly.

Short circuiting
In larger installations, that are using multiple modular boilers, adequate hydraulic isolation between the boilers is necessary to prevent short circuiting. This may occur where heat generation from the lead boiler is lost via the other dormant boilers that are not currently firing.

Short cycling
It is important that boiler capacity matches demand. Careful consideration needs to be made to ensure that the boiler minimum firing capacity does exceed the current system load. If not short cycling may be caused and energy can be wasted.

Although outside the direct remit of this Guide, energy managers should be aware of and encourage appropriate designs, systems, technologies and user behaviour with respect to management of hot and cold water consumption.

Water, like energy, is a valuable resource. In larger installations especially, myriad pumping systems will be used to move water around. Using less water overall is kinder to the environment — it also requires less energy to get it to the point of use.

C.7 Ventilation and air conditioning

Adequate levels of ventilation within buildings are important to provide a healthy interior environment. Depending on the complexity of the installations, ventilation will either natural ventilation using openings in the building fabric (windows etc.), or mechanically driven forced ventilation.

Allowing natural ventilation is clearly a low energy solution, but allowing windows to open in a space that is simultaneously being heated will waste energy. Some environments though, such as deep plan offices, secure buildings or high rise buildings make forced ventilation a necessity.

Installations with high atriums can make use of the natural phenomena of hot air rising to provide for an element of natural ventilation under more controlled conditions.

Older installations, using mechanical ventilation, may use separate supply and extract air handling units, with little control between the two units. More modern units allow some recovery of heat from the extract air to temper the coolness of the fresh supply air, before that in turn is heated to the correct temperature.

The European Directive for energy related products requires air conditioning units to use heat recovery. The efficiency of these units (expressed in percentage terms) is required to increase in accordance with implementation dates with 2016 and 2018 being key milestones. There are exceptions to this directive for specialist types of installation.

Some specialist installations, such as hospitals, require air handling units with additional higher standards of filtrations, especially for areas with critically ill patients. This can increase the energy demands of the respective air handling plant, both in terms of motor sizes and also for operational hours. These factors need to be fully accounted for in any energy strategy for the installation.

Designing the correct capacity for air handling systems is vital. Systems can be over engineered because of incorrect data from the prospective clients, or because of an overestimation for future use. What follows can be a tendency to put additional controls and associated technologies in place such as variable speed drives. These in turn can create harmonics on the electrical distribution system and create inefficiencies elsewhere within the building.

Maintenance is quite a critical consideration too; the action of the air movement within the ventilation ductwork does create accumulations of dust. As well as being a breeding ground for disease, the dust also reduces the efficiency of the ventilation system.

Ventilation strategy — heat recovery

Costlier, but increasingly common in new builds and refurbishment projects, is the introduction of active technology improvements to either recover heat losses in the ventilation system or assist with cooling during warmer months.

For example, all spaces within buildings require ventilation to keep them fresh and provide a healthy environment for the occupants. Ideally this should be natural ventilation; however, where mechanised ventilation is required, recovering heat from the extraction system's stale air helps to moderate the temperature of fresh cold air coming into a space.

Ventilation strategy — night purge

During the day and/or periods of occupation internal temperatures can build up due to the occupants, thermal loads (such as IT equipment, machinery, etc.) and the thermal mass of the building.

Night purge (or 'night flushing') allows windows, other passive ventilation openings and heating and vent systems that may normally be closed during the day to open at night and be used either by stacking or extraction to remove the internal heat that has built up, thus removing or reducing the need for mechanical cooling.

Ventilation strategy — free cooling

Free cooling is an approach to lowering the air temperature in a building or data centre by using naturally cool air or water instead of mechanical cooling. In actual practice, free cooling is not entirely free, because pumps, fans and other air/water-handling equipment is needed.

C.8 Refrigeration

The process of refrigeration uses compression of a refrigerant gas to absorb heat from a controlled environment and transfer it to another environment. Here the rejected heat can be used for either another purpose (air source heat pump for instance) or dumped into the atmosphere, either within the building or externally.

Obvious examples of use include domestic refrigerators and freezers. Commercial installations such as supermarkets, shops, restaurants and hotels are all large users of refrigeration systems. They are also used within air-conditioning to assist with cooler air in warmer months. There is also heavy use of refrigeration in large scale industry, especially the food processing industry, and pharmaceuticals.

Efficient use of energy for applications that require cool environments will depend on various factors including the thermal properties of the type of refrigerant gas used, the pressure at which the gas has to operate, the length of pipe runs and the necessary motor size to drive the system.

Other factors that will influence the type of system selected include cost considerations for installation and maintenance and the environmental impact of the selected refrigerant gas.

Some supermarkets are leading with innovative technologies and design philosophies to improve the day to day operational efficiency of refrigerators and freezers and also to improve the recovery of the rejected heat for other uses within the store.

Like hot water, correct temperature set points for refrigerators and freezers are quite critical for successful operation and to prevent irreparable damage to products. In hospitals entire stocks of clinical products can be ruined if the unit does not operate satisfactorily. In commercial operations, large quantities of stock can be ruined. Such considerations may override the need to save a few watts of energy.

C.9 Motors

Motors represent a large proportion of electrical energy consumption in the UK.

Examples of applications in the built environment may include motors driving:

(a) fans for ventilation and air-conditioning systems;
(b) pumps for refrigeration and chilling applications;
(c) air compressors;
(d) pumps for hot water and heating applications;
(e) pumps for renewable technologies such as heat pumps;
(f) lifts, escalators and moving walkways;
(g) hoists and cranes;
(h) packaging and process conveying;

(i) production machines such as paper making, textiles, metal and woodworking lathes; and

(j) domestic central heating and hot water systems.

The following extract illustration from Carbon Trust document ECA764 *Motors and drives* shows the typical losses from motor systems.

▼ **Figure C9.1** Typical losses for motor and drive systems

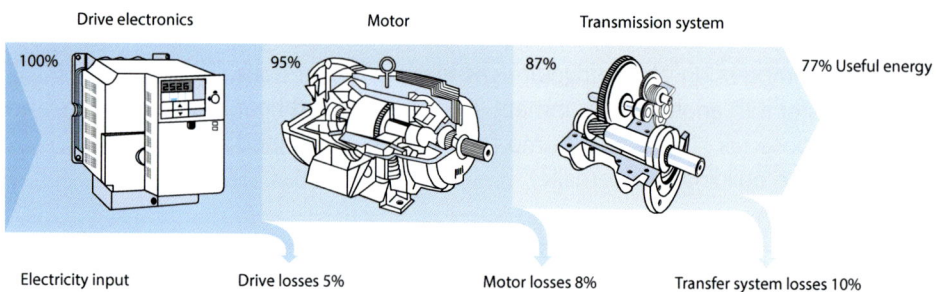

From an energy efficiency point of view, centrifugal pumps and fans have the greatest potential for energy savings as these are variable torque loads. Use of appropriately rated variable speed drives (VSD) can mean that a 20 % speed reduction may result in a 50 % energy reduction. It should also be noted that the application of VSD needs to be carefully considered against the impact of associated harmonics on the wider electrical distribution system.

There have been significant technological advances on motor efficiencies in recent years. These advances have been partially led by legislation raising the bar on acceptable efficiency standards.

As an example, IEC 60034-30 (*Rotating electrical machines – Part 30: Efficiency classes of single-speed, three-phase cage-induction motor*) classifies efficiency for these types of motors as one of three categories:

(a) IE1 standard efficiency;

(b) IE2 high efficiency; and

(c) IE3 premium efficiency.

Another category is under development – IE4 super premium efficiency.

▼ **Figure C9.2** Efficiency bands according to motor size for 4 pole AC induction motors

New IEC -Efficiency Classes in accordance with IEC 60034-30

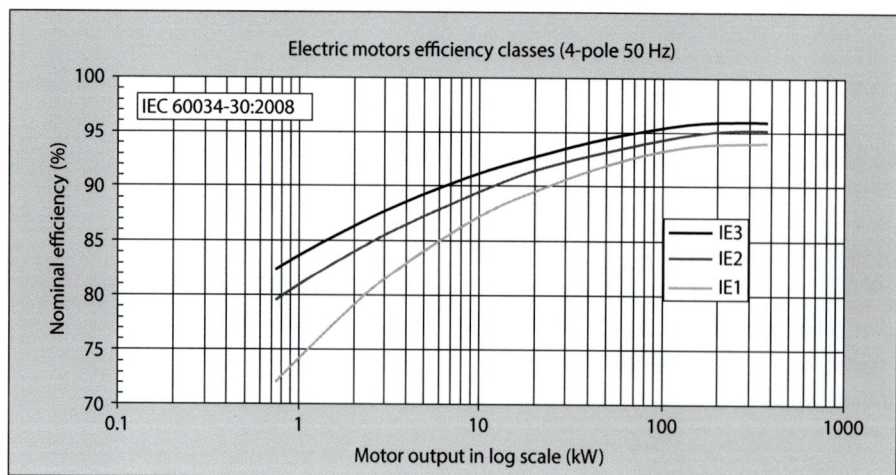

These minimum efficiency requirements have been introduced in three stages and are mandatory for AC induction motors being placed on the market or put into service.

There are exclusions to these requirements if motors are used in specific applications such as when immersed in liquids or used in explosive atmospheres or used in brake systems.

In general, the following implementation dates apply:

Mandatory stage	Implementation date	Comments
Stage 1	16th June 2011	Motors must meet the IE2 efficiency level.
Stage 2	1st January 2015	Motors, with a rated output of 7.5 kW - 375 kW, must meet either: • the IE3 efficiency level; or • the IE2 level and be equipped with a variable speed drive.
Stage 3	1st January 2017	Motors, with a rated output of 0.75 kW - 375 kW, must meet either: • the IE3 efficiency level; or • the IE2 level and be equipped with a variable speed drive.

Further study

IET	*Design Guide to Energy Efficient Electrical Installations*
Carbon Trust	ECA 764 *Motors and Drives*

C.10 Alternative/renewable energy

Criteria for building loads		Commentary
Requirements for renewable energy	Why?	Whilst optional, and a considerable investment at the present time, renewable electrical energy systems have increasingly been installed to fulfil ecological aspirations or provide a potential revenue stream.
	How?	Going forwards, the concept of some form of energy storage derived from local generation will become important for any electrical installation. The spectre of winter power cuts makes a renewable infrastructure under local control more attractive from a resilience perspective. An energy strategy that includes embedded renewable power generation in buildings, combined with demand management, appropriate controls and electrical storage should improve security of supply locally, and also improve the resilience of the wider electricity supply and distribution grid.

Further notes

Dependence on grid-connected energy supplies are likely to remain for decades to come but the energy input to an installation can be complemented by local renewables and off-grid supplies. There may also be some energy recovery from systems such as the ventilation heat exchangers or battery storage.

By utilising an alternative energy source, such as heat pumps from ground or air source, or local photovoltaic panels and wind generation, some of the local energy demand can be met in a more sustainable way. This reduces reliance on grid-connected energy supplies.

It is worth noting that heat pumps and other similar technologies are often reliant on electrically driven motors (compressors) and electronic control. Some considerations around this are:

(a) a heat pump is often capable of achieving a coefficient of performance (CoP) of 3 or greater — for every 1 kW of power drawn from the electrical grid, 3 kW of heat energy can be put into the installation;

(b) from a cost perspective, electrical energy per kWh from the grid is typically about 3 to 5 times greater than the equivalent in gas; and

(c) the cost to the end user does not vary from a conventional gas connected heat system to an electrically driven heat pump system although the carbon emissions will be less.

To offset this perhaps heat pump implementation should be provided in conjunction with solar harvested electrical supplies to provide for a real off-grid solution. Integration of photovoltaic panels within the building façade and roof profile can enhance the aesthetic appeal of a building and also its ability to be self-sustaining from an electrical energy perspective. Introduction of energy storage adds to the resilience of this kind of solution, so that the energy is used when it is most needed.

A holistic approach to energy management properly considers the causes and effects across a much wider range of issues. Business cases for modern developments should be considering these measures and their early inclusion in the design process is to be encouraged.

▼ **Figure C10.1** Introduction of renewable energy

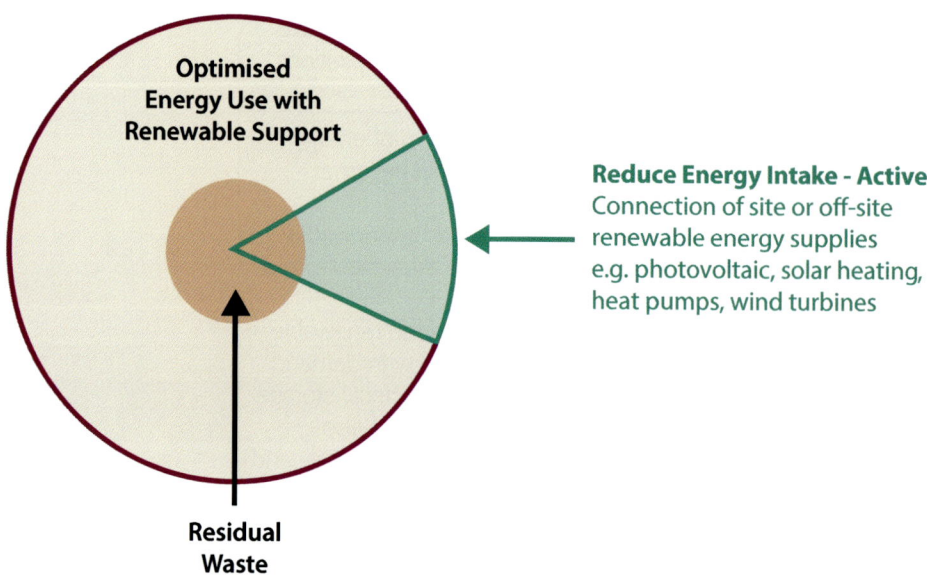

APPENDIX D

Standards and references

D.1 The context of energy management

The moral arguments for energy management are set in context by the legislative background with various national and international regulations and standards.

D.2 European directives

Within the European Union (EU), various standards and legislation have been adopted to improve energy efficiency, reduce carbon emissions and increase the use of renewable energy sources. These apply to individual systems in buildings, to energy using products and to a holistic whole building approach to energy management and efficiency:

(a) The Energy Efficiency Directive (2012/27/EU):
This directive requires all EU countries to use energy more efficiently through all stages of the energy chain from production to final consumption. It sets a target of 20 % reduction in emissions by 2020.

(b) The Energy Performance of Buildings Directive (2010/31/EU):
This directive builds on legacy documents to improve the energy performance of buildings in the EU and covers energy used for major building service infrastructure such as heating, hot water, cooling, ventilation and lighting. It also provides measures for the building envelope.

(c) The Eco-design Directive (2009/125/EC):
This directive looks at the use of energy by particular products and provides a framework to oblige manufacturers to reduce energy use of those products at the design stage. It also enforces environmental considerations during and after use.

(d) The Renewable Energy Directive (2009/28/EC):
This directive promotes the use of renewable energy and sets targets with the aim of making renewable energy sources account for 20 % of EU energy by 2020. The transport sector is highlighted to achieve 10 % use of renewable energy sources.

D.3 UK regulations

Within the UK these European Directives are augmented by legislation known as Statutory Instruments. Some of the more recent documents include the following:

Title
The Energy Savings Opportunity Scheme (Amendment) Regulations
The Energy Performance of Buildings (England and Wales)
The Pollution Prevention and Control (Designation of Energy Efficiency Directive) (England and Wales)
The Ecodesign for Energy-Related Products and Energy Information (Amendment) Regulations
The Energy Efficiency (Building Renovation and Reporting) Regulations
The Energy Efficiency (Eligible Buildings) Regulations
The Promotion of the Use of Energy from Renewable Sources (Amendment) Regulations

The Energy Efficiency (Private Rented Property) (England and Wales) Regulations 2015

Energy managers should note that the above regulations have implemented Minimum Energy Efficiency Standards (MEES).

These standards were introduced by the government in the Energy Act 2011.

These changes apply to both residential and commercial properties and come into force on 1st April 2018.

D.4 ISO 50000 series of documents

As the lead document ISO 50001 provides a framework, within Section 4 of that document, which correlates with the PDCA management cycle. The four stages are set as a continuous process, which should then repeat on itself at regular intervals. New information from regular monitoring or as outside influences can be added as the basis for controlled changes and further improvements.

Annex B of ISO 50001 provides a table that shows how the format of the document, and hence its processes, is aligned to other closely related standards, such as ISO 9001 and ISO 14001.

The main operational headings within the document are shown in the table below. The 'Headline activity' column of the table also shows how these headings align to the PDCA cycle.

2) Plan

Section	Heading	Sub heading	Headline activity	Key question
4.2	Management responsibility		Plan	
4.2.1		Top management		Who?
4.2.2		Management representative		Who?
4.3	Energy policy		Plan	
4.4	Energy planning		Plan	
4.4.1		General		Why?
4.4.2		Legal requirements and other requirements		Why?
4.4.3		Energy review		How?
4.4.4		Energy baseline		When?
4.4.5		Energy performance indicators		Which?
4.4.6		Energy objectives, energy targets and energy management action plans		What?

3) Do

Section	Heading	Sub heading	Headline activity	Key question
4.5	Implementation and operation		Do	
4.5.1		General		Why?
4.5.2		Competence, training and awareness		Who?
4.5.3		Communication		How?
4.5.4		Documentation		How?
4.5.5		Operational control		How?
4.5.6		Design		Which?
4.5.7		Procurement of energy services, products, equipment and energy		What?

4) Check

Section	Heading	Sub heading	Headline activity	Key question
4.6	Checking		Check	
4.6.1		Monitoring, measurement and analysis		What?
4.6.2		Evaluation of compliance with legal requirements and other requirements		Which?
4.6.3		Internal audit of the EnMS		Why?
4.6.4		Non-conformities, correction, corrective and preventive action		Why?
4.6.5		Control of records		How?

5) Act

Section	Heading	Sub heading	Headline activity	Key question
4.7	Management review		Act	
4.7.1		General		Why?
4.7.2		Input to management review		Which?
4.7.3		Output from management review		What?

The international ISO 50000 series of documents are managed under the ISO/TC 242 Energy Management.

	Standard and/or project	
1	ISO 50001	Energy management systems — Requirements with guidance for use
2	ISO 50002	Energy audits — Requirements with guidance for use
3	ISO 50003	EMS — Requirements for bodies providing audit and certification of energy management systems
4	ISO 50004	EMS — Guidance for the implementation, maintenance and improvement of an energy management system
5	ISO 50006	EMS — Measuring energy performance using energy baselines (EnB) and energy performance indicators (EnPI) —General principles and guidance
6	ISO 50015	EMS — Measurement and verification of energy performance of organizations — General principles and guidance

These documents, led by ISO 50001, provide a structured framework to compiling an energy management system that can be tailored to any installation. It should be noted that further titles are being developed in this series of documents.

Associated matrix for similar activities

Management	Outline and detailed design	Installation and commissioning	Operational standard	Comments
Safety	Design risk assessments	Health and safety Method statements Risk assessments	ISO 9001	Health & safety policy Management of residual risks Management of maintenance risks
Environment	Sustainable energy sources			Environmental policy Local procurement
	Sustainable disposal	WEEE	ISO 14001	Recycle/ sustainable disposal
Energy	Low energy products	Controls Commissioning Measurement Auditing	ISO 50001	Energy policy Energy strategy Management of energy reduction/ optimisation

D.5 Additional references

Reference information (legislation/regulations/standards/guidance)

Reference	Title/comments
PD CEN/CLC TR 16103:2010	Energy management and energy efficiency. Glossary of terms A Technical Report defining key terms commonly used in energy management and energy efficiency
PD ISO/TR 16344:2012	Energy performance of buildings. Common terms, definitions and symbols for the overall energy performance rating and certification
PD CEN/TS 16628:2014	Energy Performance of Buildings. Basic Principles for the set of EPB standards
PD CEN/TS 16629:2014	Energy Performance of Buildings. Detailed Technical Rules for the set of EPB-standards
ISO 23045:2008 Ed 1	Building environment design. Guidelines to assess energy efficiency of new buildings
BS EN 15900:2010	Energy efficiency services. Definitions and requirements Used as a reference document for appropriate qualification, accreditation and/or certification schemes for providers of energy efficiency services, as mentioned in Article 8 of Directive 2006/32/EC
PAS 2030:2014	Online Single Licence Improving the energy efficiency of existing buildings. Specification for installation process, process management and service provision

Reference	Title/comments
PAS 2031:2015	Certification of energy efficiency measure (EEM) installation services applicable to certification bodies that are providing, or intending to provide, conformity evaluation services in respect of PAS 2030, Improving the energy efficiency of existing buildings — Specification for installation process, process management and service provision.
PAS 2080:2016	Carbon management in infrastructure PAS 2080 provides a common framework for all infrastructure sectors and value chain members on how to manage whole life carbon when delivering infrastructure assets and programmes of work.
BS EN 16247-1:2012	Energy audits. General requirements
BS EN 16247-2:2014	Energy audits. Buildings
BS EN 16247-3:2014	Energy audits. Processes
BS EN 16247-4:2014	Energy audits. Transport
BS EN 16247-5:2015	Energy audits. Competence of energy auditors
BIP 2187:2011	Energy management principles and practice (second edition)
BS EN 15232:2012	Energy performance of buildings — Impact of Building Automation, Controls and Building Management
BS EN 16231:2012	Energy efficiency benchmarking methodology
BS EN 15500 (2008)	Control for heating, ventilating and air-conditioning applications — Electronic individual zone control equipment
BS 8206-2 (2008)	Code of Practice for daylighting
BS EN 12464-1(2011)	Light and lighting. Lighting of work places — Indoor work places
BS EN 15193 (2007)	Energy performance of buildings, Energy requirements for lighting
IET Standards	Code of Practice for Grid Connected Solar Photovoltaic Systems
BS EN ISO 25745-1 (2012)	Energy Performance of Lifts, Escalators and Moving Walks — Part 1: Energy Measurement and Verification provides an agreed international method of energy measurement for lifts
BS EN ISO 25745-2 (2014)	Energy performance of lifts, escalators and moving walks — Part 2: Energy calculation and classification for lifts elevators
BS EN ISO 25745-3 (2014)	Energy performance of lifts, escalators and moving walks — Part 3: Energy calculation and classification for escalators and moving walks
CIBSE Guide D	Transportation systems in buildings
CTG 050 Carbon Trust	Power play — Applying renewable energy technologies to existing buildings
CTV 038 Carbon Trust	Low Carbon Refurbishment of Buildings — A guide to achieving carbon savings from refurbishment of non-domestic buildings

Reference	Title/comments
PD CEN/CLC/TR 16567:2013	Energy Efficiency Obligation Schemes in Europe. Overview and analysis of main features and possibilities for harmonisation
BIP 2221:2013	Implementing and Improving an Energy Management System: How to Meet the Requirements of ISO 50001
Directive 89/106/EEC	on the approximation of laws, regulations and administrative provisions of the Member States relating to construction products
Directive 92/42/EEC	on efficiency requirements for new hot-water boilers fired with liquid or gaseous fuels
Directive 92/75/EEC	on the indication by labelling and standard product information of the consumption of energy and other resources by household appliances
Directive 2004/8/EC	on the promotion of cogeneration based on a useful heat demand in the internal energy market
Directive 2006/32/EC	on energy end-use efficiency and energy services
Directive 2005/32/EC	establishing a framework for the setting of eco-design requirements for energy-using products
Directive 2009/28/EC	on the promotion of the use of energy from renewable sources
Directive 2010/30/EU	on the indication by labelling and standard product information of the consumption of energy and other resources by energy-related products
Directive 2010/31/EU	on the energy performance of buildings (recast)
BRE IP 1/14	Lewry, Andrew J. Understanding the choices for building controls IHS BRE Press, 2014.
BRE IP 1/15	Lewry, Andrew J. Bridging the performance gap — Understanding the predicted and actual energy use of buildings. IHS BRE Press, 2015

D.6 Commentary on procurement references

Review and use of the following best practice procurement standards is recommended:

(a) EU Green Public Procurement (GPP) criteria [1]

This relies on having clear, verifiable, justifiable and ambitious environmental criteria for products and services, based on a life-cycle approach and scientific evidence base.

(b) Eco-design Directive (2009/125/EC) [2]

This provides EU-wide rules for improving the environmental performance of energy related products (ERPs) through eco-design.

It requires minimum energy performance criteria from a range of technologies used in buildings.

(c) UK's Enhanced Capital Allowance (ECA) scheme

The ECA scheme supports investments in certain energy saving equipment, against the taxable profits of the period of investment.

For a product to be eligible for ECAs it must meet specific energy saving eligibility criteria. The scheme is underpinned by the Energy Technology List (ETL), which currently has criteria for 17 technology areas and 60 sub-technologies.

(d) UK's Government Buying Standards (formerly known as Buy Sustainable Quick Wins)

These contain official specifications that all UK government buyers must follow when procuring a range of products.

Energy models

E.1 Energy models

Understanding what energy model an installation has and what alternatives are technically feasible can help to inform and influence an energy management strategy.

▼ **Figure E.1**   Basic model of energy resourcing

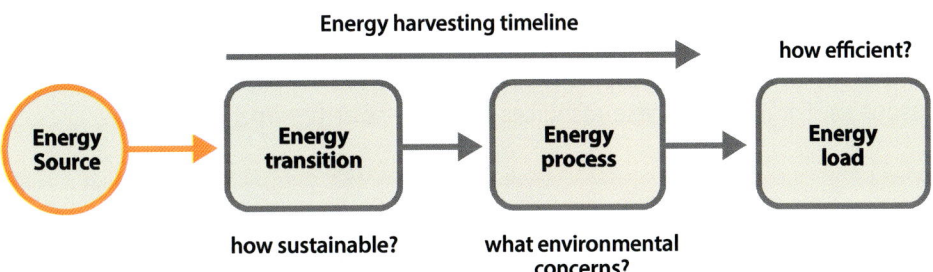

An energy model will vary from organisation to organisation. It may even vary from one form of energy to another, for example:

(a) gas is likely to be connected to the local grid or brought in by tankers; and

(b) electricity is predominantly grid-connected, but increasingly this will be supplemented by local solar generation – in rare cases an electrical installation may be completely off-grid.

Use of CHP is increasingly used to increase local energy efficiency by providing both heat and power; however this is still driven by fossil fuels of one kind or another. Examples of energy models can be summed up in the following simplified models.

E.2 Traditional model

With the traditional model the primary energy resource is fossil fuels laid down millions of years ago from plants and animals. This has been converted to oil and coal through natural processes that have occurred over a significant period of time and followed with extraction by modern society through industrial processes to provide energy for homes and industry.

The energy extracted is often high grade and easily converted, however the processes that consume the energy at the point of use have often been inefficient and wasteful. Pollution of the environment is another issue with this model. Replenishment of these resources once expended is not feasible.

▼ **Figure E.2** Traditional model of energy resourcing

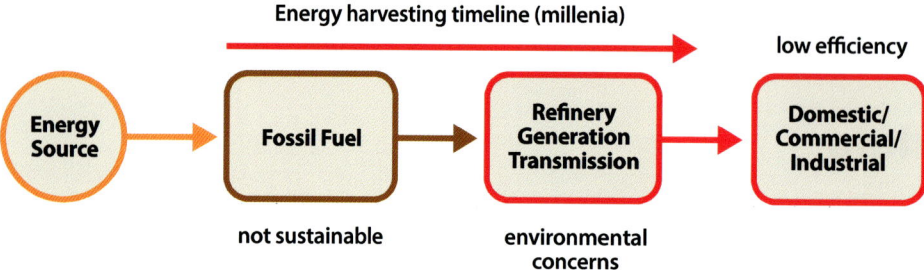

E.3 Contemporary model

With the contemporary model, the principal background supplies remain as per the traditional model. There is still an element of pollution and impact on the environment. However, these circumstances will be increasingly supplemented by alternative supplies from more sustainable resources. For instance these resources could be solar (thermal or photovoltaic) or wind.

For an installation that relies purely on these sustainable resources, resilience is potentially an issue. Without the wind, the turbines will not generate and without the sun, the solar sources will not work either. Some element of energy storage may be necessary. Pollution caused by continued use of fossil fuels is still a problem, as is replenishment of supplies. However, mitigations from the use of alternative supplies partially alleviates the problem.

The model is still dependent on fossil fuel based products to provide more sustainable energy. Heat recovery is increasingly adopted in order to increase system efficiency levels but the energy recovered is low grade and possibly difficult to manage.

▼ **Figure E.3** Contemporary model of energy resourcing

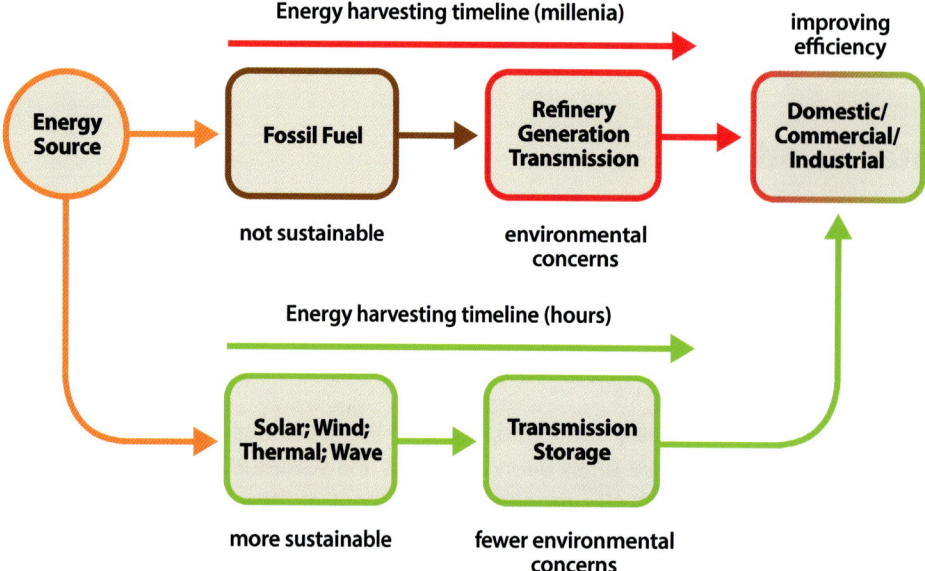

E.4 Future model

With the future model, there will be new technologies that have not yet come to market or perhaps existing technologies with new and better design philosophies. These will provide energy from completely renewable supplies under all circumstances.

This model will cut out the 'middle man' of fossil fuel based supplies to provide a much more sustainable operation. The problem of replenishment of energy supplies will also be solved as the issue of fossil fuel production is bypassed and energy harvesting directly from the effects of solar energy become the norm rather than an expensive 'greener' accessory.

Resilience issues for mechanical and electrical energy sources will be overcome by effective energy storage. Learning to work with the sun and with the limitations of harvesting energy in daylight hours will represent a challenge to the energy industry.

The model provides products from a completely sustainable supply chain predominantly based on local resources. It also deals efficiently with low grade heat recovery.

▼ **Figure E.4** Future model of energy resourcing

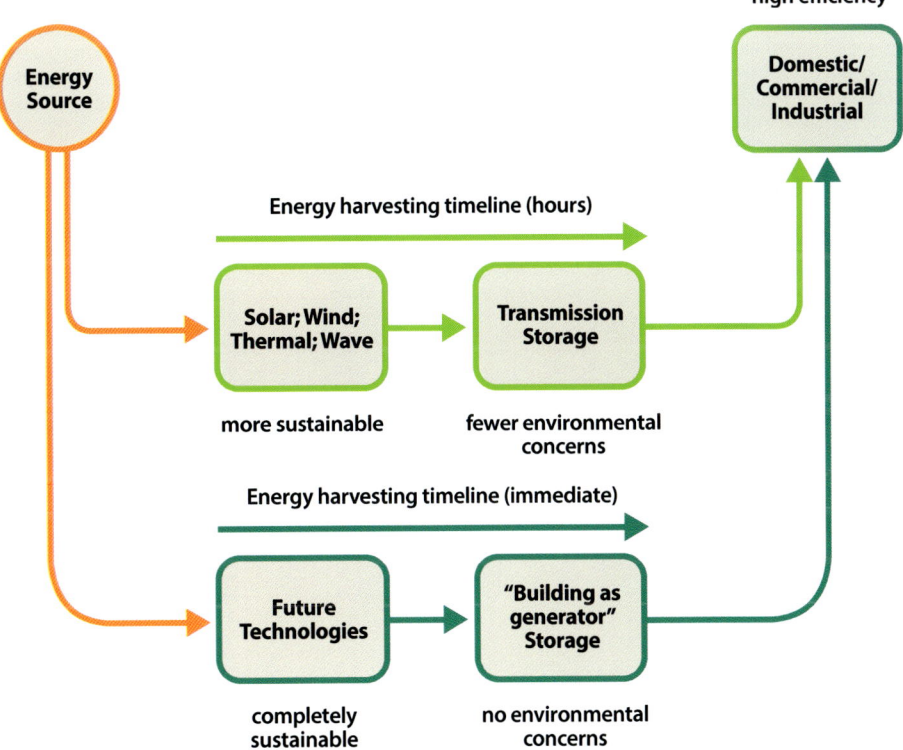

Future energy management techniques

F.1 General

Transition engineering (TE) is an emerging field that challenges existing short to medium term philosophy and changes the focus and dynamic to a longer term view. Within the context of energy management, TE encourages the energy manager to look for the solution first and not necessarily at the problem.

Using existing philosophy and change management under ISO 50001 and similar mechanisms, considerable reductions in energy consumption can be achieved. However, this is reliant on making efficiencies in the existing use of fossil based fuels and grid connections. It is supported by legislation and current political will to lower emissions from fossil fuels. However, there are often commercial barriers and political hurdles which create an artificial environment where only short term energy management policies are implemented.

Typical human thinking has always been to deliver more energy with more resources. Efficiency and challenges to existing technologies have not always been considered too much of a priority. There has to be a change of mindset to deliver enough energy with fewer resources. Much more long term thinking is really required to re-inform and direct short and medium term activities so that they are more productive and with a more coherent strategy.

F.2 Seven steps of transition engineering

Using a seven step process TE takes stock of where we are now in the context of energy supply and demand, recognises how we got here, and where the path may lead us on a similar trajectory. It then looks at what we would do with a less traditional energy resource. Using a variety of tools it works back from the solution to recognise trigger events for change and implements change projects to introduce new technologies and operational philosophies to bring us closer to the solution.

Much has been written about peak oil production and the concerns that it raises for the way modern society now lives. More has been written on the effects of fossil fuel use, the consequent impact on the environment and the effects that would have on the way society would be forced to live. Those are the problems associated with existing technology and the philosophies of the contemporary world.

"The solution to the problem of climate change is to leave the carbon from the age of dinosaurs safely sequestered in the ground. The solution to the problem of the depletion of conventional oil supply is to use less oil. Keep it simple. Think about the solution not the problem." (Source: TE website) http://transitionengineering.co.nz

The question to be posed is: what engineering systems do we need in place to support our modern world based on less or no oil?

The process of transition engineering designs and implements projects that manage the risks of unsustainable resources and the associated environmental impacts. TE involves detailed research and analysis and works with local organisations to ensure satisfactory solutions. It also examines economic balance and social consistency. The projects are then carried out 'from the ground up' informed by local knowledge.

Other goals include provision of local jobs, improvements in local resilience and better potential for adaptation in local approaches. There is an acceptance that use of less (and ideally no) fossil fuel is business as usual.

▼ **Figure F.1   Seven steps of transition engineering**

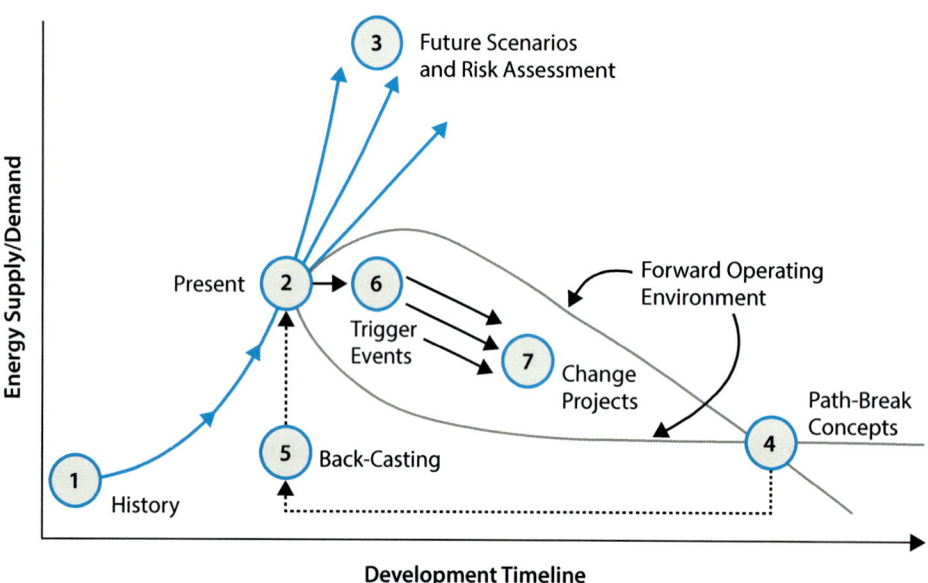

F.3 The process of transition engineering

The transition engineering resource pages provide the following explanations surrounding the seven step process.

(a) Past trends
   i   Drive the need to understanding past activities and performance.
   ii   What are the past trends?
   iii   How were things in the past, and how have they changed?
   iv   What are the resources that have always been important, and what are the roots of our shared values?
   v   Data gathered by government is analysed, but community records and oral histories are also important information to be gathered by discussion with long-term residents.

(b) Current situation
   i   Fulfils the need to understand the current situation.
   ii   Government statistics define the current state of the community.
   iii   The perceptions and attitudes of residents are also important and are best understood by discussions with a wide range of individuals and groups.
   iv   The regional assets and resources are also understood through interviews.

(c) Future scenarios
  i   Forecasts various future scenarios based on existing patterns of use and conventional engineering systems.
  ii If a trend is unsustainable, then it will change.
  iii Future trends can be explored using simple mathematical models and a range of assumptions.
  iv A future timeline is developed that sets the framework for development of targets for changes in unsustainable trends.

(d) Path-break concepts in the forward operating environment
  i   Requires creative and imaginative thinking.
  ii Concepts are generated by collaborative and multidisciplinary brainstorming process and have multiple benefits and are mutually beneficial for the sectors involved.
  iii Enterprises or systems, at a future time, are conceived to meet the essential needs of people and operate sustainably.
  iv Historical issues and current problems do not hinder the concept generation.
  v   Strategic analysis of complex systems method is used to quantitatively explore the possible concepts in preparation for a 'Transition>Scape' workshop.

(e) Back-casting
  1   Analysis of where the operating environment should be compared to where it is now - what are the gaps and how can they be changed?
  2   Resources in the local community and region are evaluated in relation to each concept.
  3   Back-casting evaluates the resourcefulness of particular people, business opportunities, known technologies, innovative ideas and synergies.
  4   Barriers to the development of the path-break concepts are also explored.

(f) Triggers
  i   Use of a change management tool already deployed in the fields of innovation, sustainability and strategy. These can also be events that are a catalyst for change.
  ii The trigger project is a pilot-scale activity that brings together the resources and talents in the community and demonstrates the beneficial outcomes.
  iii The trigger project is also an opportunity to work out the problems and make adjustments to the project.
  iv Triggers can also be external disruptions, like a fuel shortage, or failures of the existing systems, like a high death toll amongst cyclists, or poor public health outcomes, like respiratory disease from cold, mouldy conditions in homes.

(g) Transition projects
  i   Programme planning to realise goals.
  ii Transition projects are change projects led by communities and organizations, which will probably lead to adaptation of measures by government later.
  iii Transition projects involving infrastructure, transport, energy and buildings will require professional work to be commissioned.

(Source: TE website)

INDEX

N no entries

O

objectives	2.6
operational energy management	A1.4
operational parameters	2.6; 2.10

P

performance: *see* energy performance	
'Plan, Do, Check, Act' (PDCA)	1.6
policies	2.6; 2.8; A1.1
procurement of energy	2.6; 2.9; A2.1; D.6

Q no entries

R

refrigeration	C.8
regulations	D.3
renewable energy	2.5.1; C.10
risk assessment	A1.3
roles and responsibilities	2.6; 2.9; A2.2

S

self-assessment questions	Appendix B
sensors	2.5.2; 2.9; C.3
standards	1.6; 1.7; D.4; D.5
strategies	2.6; 2.8; A1.2

T

tariffs	2.9
technical guidance	2.4; D.5
thermal performance	A1.2; C.1
transition engineering (TE)	Appendix F
trends in energy use	2.11

U

V

W

X, Y, Z no entries

IET Standards

Influence the future of Standards

Working in a fast-paced, rapidly changing industry has its frustrations. A lack of professional standards and guidance increases risk, and hinders the ability to embrace innovation.

The IET uses its wealth of knowledge and experience to bring about standards that:

- Solve common working problems
- Make meeting legislative requirements simple
- Give practical guidance for practising engineers

To create the best possible guidance, we need you.

Get involved as a member of a publication committee, take on authorship of a book or simply give us feedback on a draft publication.

Find out more at:

www.**theiet**.org/setting-standards

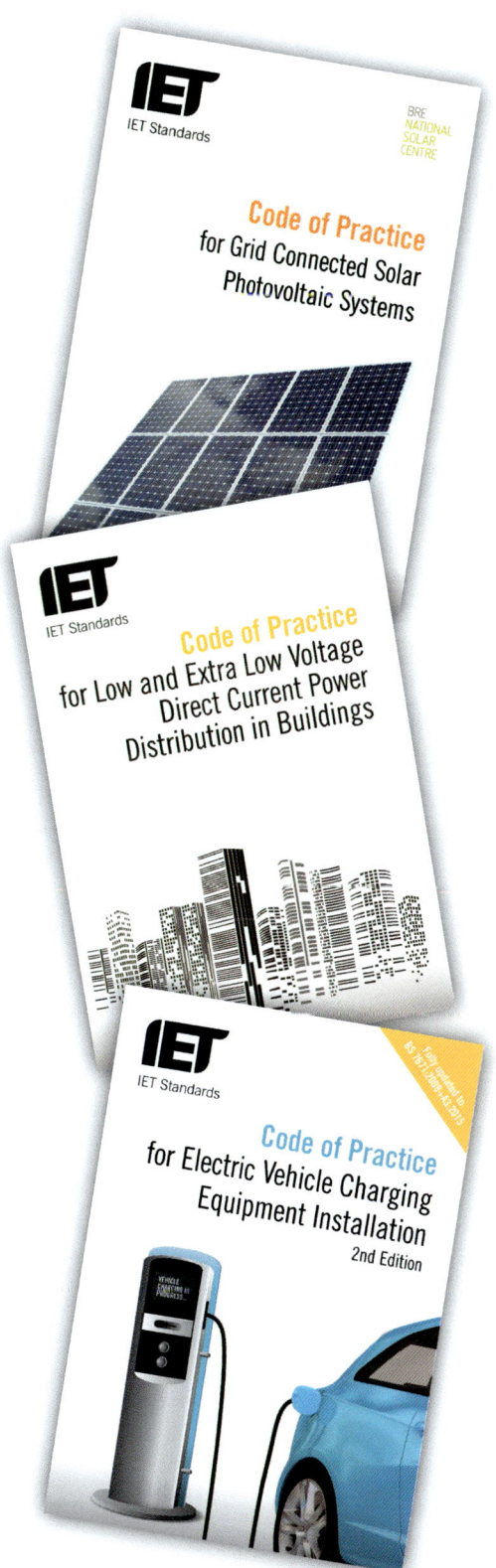

Electrical **excellence**

Industry-leading standards

IET Standards works with industry-leading bodies and experts to publish a range of Codes of Practice and guidance materials for professional engineers, using its expertise to achieve consensus on best practice in both emerging and established fields. Titles related to this Guide include the *Guide to Metering Systems* and *Code of Practice for Electrical Energy Storage Systems*.

See these titles and more at:

www.theiet.org/standards

ELECTRICAL STANDARDS +

Constantly up-to-date digital subscriptions

Our expert content is also available through a digital subscription to the IET's Electrical Standards Plus platform. A subscription always provides the newest content, giving peace of mind that you are always working to the latest guidance.

Going digital gives you greater flexibility when working with the Wiring Regulations, Guidance Notes and the IET's expert Codes of Practice available for electrical engineers. The intuitive search function instantly serves results from across all books in your package. You can also access the content on your desktop, laptop or tablet, making it easy to take the content out on site or read on the move.

Find out more about our subscription packages and choose one to suit you at:

www.theiet.org/digital-regs